A Century of Eugenics in America

BIOETHICS AND THE HUMANITIES

ERIC M. MESLIN AND RICHARD B. MILLER, EDITORS

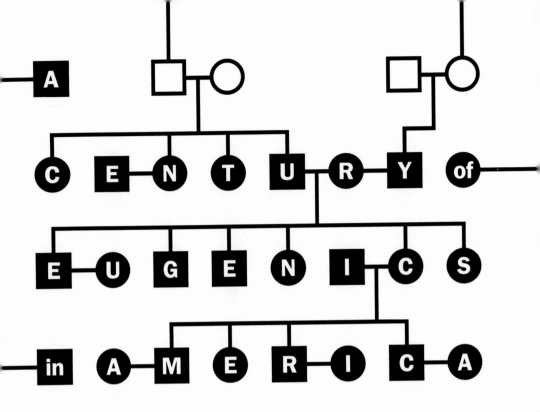

A CENTURY of EUGENICS in AMERICA

From the Indiana Experiment
to the Human Genome Era

EDITED BY

PAUL A. LOMBARDO

INDIANA UNIVERSITY PRESS

Bloomington & Indianapolis

This book is a publication of

Indiana University Press
601 North Morton Street
Bloomington, IN 47404-3797 USA

iupress.indiana.edu

Telephone orders   800-842-6796
Fax orders   812-855-7931
Orders by e-mail   iuporder@indiana.edu

Manufactured in the United States of
America

Library of Congress Cataloging-in-Publi-
cation Data

A century of eugenics in America : from the
Indiana experiment to the human genome
era / edited by Paul A. Lombardo.
       p. ; cm. — (Bioethics and the
humanities)
   Includes bibliographical references and
index.
   ISBN 978-0-253-35574-4 (cloth : alk.
paper) — ISBN 978-0-253-22269-5 (pbk. :
alk. paper) 1. Eugenics—United States—
History—20th century. I. Lombardo, Paul
A. II. Series: Bioethics and the humanities.
   [DNLM: 1. Human Genome Project.
2. Eugenics—history—United States.
3. Eugenics—legislation & jurisprudence—United States. 4. History, 20th
Century—United States. 5. History, 21st
Century—United States. 6. Sterilization,
Involuntary—history—United States.
7. Sterilization, Involuntary—legislation
& jurisprudence—United States. HQ
755.5.U5 C397 2010]
   HQ755.5.U5.C46 2010
   363.9'2—dc22
                                    2010020427

2 3 4 5   16 15 14 13 12

FOR CONNI, CHRIS, AND CLARE

CONTENTS

In 1907, Indiana passed the first involuntary sterilization law in the world based on the theory of eugenics. In time more than 30 states and a dozen foreign countries followed Indiana's lead in passing sterilization laws; those and other laws restricting immigration and regulating marriage on "eugenic" grounds were still in effect in the United States as late as the 1970s.

The centennial of Indiana's pioneering enactment provided an opportune time to evaluate the historical significance of eugenics in America. On April 12, 2007, after more than two and a half years of planning, a group that included scholars, state officials, and members of the public assembled in Indianapolis for the culmination of the Indiana Eugenics Legacy Project. The project was designed to advance historical research on eugenics, to deepen our understanding of the varied ways "eugenics" was expressed intellectually, legally, and socially, and to help draw lessons from history for current policy makers.

The project included several specific events. Foremost among them was a public symposium held to mark the eugenic centennial. On that occasion the Indiana State Library launched an exhibit on the history of eugenics in Indiana as scholars engaged in panel discussions on the implications of eugenic policies. Professor Daniel Kevles, historian and author of *In the Name of Eugenics* (1985), provided a keynote lecture, as did Joe Palca, science journalist for National Public Radio. The highlight of the centenary activities was the unveiling of a historical marker that now stands on the grounds of the State Library, explaining Indiana's role in the eugenics movement.

The Indiana secretary of health appeared on behalf of the governor to condemn past eugenic abuses, adding her voice to that of the Indiana legislature, which adopted a formal resolution decrying its role in eugenic history. The Indiana Supreme Court held a conference on eugenic history in its own courtroom. The coalescence of all three branches of state government to reflect on and repudiate the abuses that took place as part of America's eugenic past was unprecedented. Of particular significance was the appearance of Jamie Coleman, a woman whose challenge to the legality of her own involuntary sterilization in Indiana reached the United States Supreme Court in the 1978 case of *Stump v. Sparkman*. She attended the symposium and unveiled the historical marker.

The project also generated educational activities. In addition to the Supreme Court seminar, a graduate school course on eugenics was held at Indiana University–Purdue University Indianapolis. Articles were published in the *Indiana Magazine of History* and *Traces of Indiana and Midwestern History,* and several presentations were given by project participants at state and national meetings such as the Indiana Association of Historians, the American Public Health Association, and the American Society of Law, Medicine, and Ethics. The exhibit that was displayed at the State Library in Indianapolis was digitized for permanent display online. By all measures, the project was a success.

In the past 25 years, scholars in the United States and several other countries have documented the wide appeal of eugenics, and students of contemporary science have used that research as a warning about potentially troubling applications of the breathtaking new discoveries in genetics. This volume, the final product of the Indiana Eugenics Legacy Project, builds on that growing literature. The Indiana eugenic centennial provided an opportunity to undertake original historical research. In contrast to the many wide-ranging scholarly and popular surveys of eugenics already available, this book was planned as an exploration of the detailed and varied history of eugenics in America at the state and local levels, beginning with its appearance in Indiana.

Authors of several recent historical works that analyzed regional and state eugenic programs were invited to participate, along with other scholars from the fields of law, genetics, and bioethics. Grant funding supported the committee of scholars to travel to Indianapolis to attend public events and to discuss their papers for a commemorative volume.

This group produced new scholarship that probed practices in Indiana, Georgia, California, Minnesota, North Carolina, and Alabama, along with other papers that explored perspectives on bioethics, law, race, and eugenics. Our goal was to create a volume that would contribute to the ongoing national discussion about the meanings of "eugenics" and how those meanings played out in specific and concrete contexts.

I want to thank several key people who were instrumental in bringing the Indiana Eugenics Legacy Project to fruition. First on the list is Eric Meslin, director of the Indiana University Center for Bioethics. When I contacted Eric in 2004 with a reminder that the centennial of the Indiana sterilization law was approaching, he enlisted the expertise of IU historian William Schneider. Bill led the Legacy Project, managing the efforts to secure funding and bringing together the wide diversity of people who eventually contributed to the project's success. We then asked Alexandra Stern of the University of Michigan to join our planning group. Her own experience as a historian with an in-depth knowledge of Indiana's eugenic past represented another welcome resource. Without the skills of Eric, Bill, and Alex, this project would not have occurred.

Judi Izuka Campbell, a research associate in the Medical Humanities & Health Studies program, provided invaluable assistance in managing project logistics. Another IU colleague, David Orentlicher, used his considerable skills as physician, lawyer, and member of the Indiana House of Representatives to shepherd the eugenics resolution through the General Assembly. Thanks to Judi and David for their contributions to the project.

Special thanks are also due to the National Institutes of Health's Human Genome Institute's Ethical, Legal, and Social Issues program, which provided funding for the 2007 symposium and support for this volume. I also want to thank Dean Steven Kaminshine of the Georgia State University College of Law for providing summer support that allowed me to complete the work of compiling this volume.

Paul A. Lombardo
April 2010

A Century of Eugenics in America

Looking Back at Eugenics

PAUL A. LOMBARDO

Eugenics. A quick internet search identifies that word as the invective *du jour* in public discourse, shorthand for everything evil. Most often those who brandish the "E" word condemn it as the nadir of "pseudo-science" and make explicit reference to Hitler and the Holocaust. And after many years of absence from public consciousness, terms like *eugenicist* are now regularly employed to skewer a political opponent, to condemn the teaching of evolution, and to oppose some feature of health care reform. The specter of eugenics is also commonly invoked to question the use of new technologies and the pursuit of science more generally. Clearly the meaning of the term *eugenics*, praised a century ago as a science made up of "fact not fad,"[1] and used to signal the study of those "hereditarily endowed with noble qualities,"[2] has undergone a sea change.

Part of the reason for the recent reemergence of *eugenics* as an almost exclusively pejorative term is the expansion of scholarship that has explained the origin of the U.S eugenics movement.[3] Many people are still shocked to hear that practices such as eugenic sterilization began in the United States long before they were taken up in totalitarian settings such as Nazi Germany. Because of the power of that historical trajectory, a linkage is assumed—both too often and too quickly—between anything "eugenical" and the rise of the Third Reich.

But historians of eugenics have been saying for decades that from the first enunciation of Galton's "brief word to express the science of improving stock,"[4] eugenics took on an ever-changing variety of meanings,[5] and that those sometimes complementary, sometimes contrasting meanings generated "competing and evolving varieties of eugenics."[6]

The essays in this book do not propose a single definition that captures the meaning of *eugenics*. They are instead attempts to describe and analyze the many ways that term was used to justify cultural shifts, social programs, and laws in the United States. There is no extended discussion of the "eugenics movement" in these essays either, both because most of the nationally prominent eugenic organizations have already been well studied and because the general focus here is not on national trends but on how those trends played out in ways that were unique and local. Eugenics took many forms, and different agendas were launched "in the name of eugenics."[7] This book traces the career of several of those agendas, with particular attention to legal activities at the state level. The first eight essays are by historians; seven of them are about features of sterilization law in one or more states. While sterilization is clearly only one expression of the group of ideas we think of as eugenics, it still draws historical attention because it was practiced so regularly in the United States for so long. It also generated legal and administrative records that are the raw material of much historical study.

The eighth historical essay is about race and the way eugenic thinking of one kind was adopted even by those who might have been victimized under "eugenics" of another stripe. Race is touched upon in several other chapters, and that is hardly surprising, since racially and ethnically discriminatory laws, from Jim Crow to genocide, represent some of the most notorious examples of the policies we understand as eugenics.

The book is completed with two essays from perspectives outside of history: law and biomedical science. They too allude to history, but also bring our inquiry up to date with reflections on the era of the human genome. The book as a whole can be broken down into four parts, as summarized below.

PART 1. THE INDIANA ORIGINS
OF EUGENIC STERILIZATION

Two essays place Indiana as first among all the states to begin the "experimental stage" of eugenics, with specific attention to Dr. Harry Clay Sharp, a prime mover in the legalization of eugenic surgery, followed by an in-depth view of the state legislative process that yielded the first sterilization law.

Elof Carlson reminds us that the roots of twentieth-century eugenics had burrowed deep even before that word was coined. Oscar McCulloch's nineteenth-century tale of the Tribe of Ishmael relied on earlier degeneracy theory to convince readers of the social costs of wandering tribes and the problem families they nurtured. McCulloch's fixation on the Ishmaelites—a group not responsive to his charitable reforms—foreshadows the later account in this book of the Bunglers, another pseudonymous clan that seemed impervious to a reformer's efforts.

Carlson surveys a broad sweep of social and cultural history, reaching back to the history of English Poor Laws for an explanation of how new ideas on the role of public philanthropy developed in turn-of-the-century Indiana. He also highlights the importance of the new technology of surgical vasectomy, which became available to Dr. Sharp when his career as prison physician was just beginning and was viewed for a time as therapeutic for the criminals to whom it was applied. Its use as a eugenic tool to isolate the seeds of criminality within those already convicted of crime was considered a value-added feature of the novel surgical technique.

Jason Lantzer moves the discussion of Indiana's 1907 sterilization law forward, tracing it to interest group politics. Rather than focusing on a grassroots movement, he details the small cadre of reformers, politicians, and physicians whose first efforts at eugenic law resulted in denying marriage licenses to the poor and disabled and eventually produced sterilization laws that in various incarnations remained in place in the state for almost seven decades. Lantzer explores the political and legal climate surrounding Indiana's sterilization laws during that period and explains how the law changed over time to fit the idiosyncratic needs of advocates. He also identifies the changing targets of Indiana sterilization law, which included criminals, the "feebleminded," the mentally ill, and people with epilepsy. Similar insights about the officials who administered sterilization laws and the changing motives they articulated are evident in later local accounts of sterilization in this volume. The next two articles provide examples of the broad reach of eugenics in American culture.

PART 2. EUGENICS AND POPULAR CULTURE

A common theme among most of these essays is the role of economic pressures, real or imagined, on the adoption of practices described as "eugenic." The work of novelist Erskine Caldwell confronted Georgians with the problems of poor families in their midst during the Great Depression. Caldwell's insistent focus on the most desperate families—described in both his fiction and his newspaper reportage—heightened attention to proposals for a sterilization law as the solution to intergenerational familial poverty. Chapter 3 connects the career of Georgia's 1937 sterilization law to debates over poverty and eugenics fueled by Caldwell's work.

The second essay in this group identifies the power of eugenic thinking in unlikely places: among the writings of W.E.B. Du Bois in the National Association for the Advancement of Colored People magazine, the *Crisis,* and other Du Bois publications in Margaret Sanger's *Birth Control Review*. Gregory Dorr and Angela Logan review the writings of Du Bois alongside other African American leaders. They find a brand of "black eugenics" that de-emphasized interracial differences and instead encouraged reproduction of quality traits among the most "fit" members of black society. They analyze a baby contest sponsored by the NAACP, started as a vehicle for raising money to combat the scourge of lynching. The contest succeeded as a funding vehicle, while simultaneously highlighting the growth of the "talented tenth" within Du Bois's own community.

PART 3. STATE STUDIES OF EUGENIC STERILIZATION

The next four chapters are case studies that further explain how eugenic sterilization law was applied in key states. Alexandra Stern details the use of eugenic sterilization in mental institutions in California and compares western practices to the application of surgery in the correctional system in Indiana. Her chapter introduces us to the lives of those most affected by sterilization law—institutionalized patients. Their stories are revealed through hospital records and other archival material containing accounts of both resistance to sterilization and acquiescence to its application by the families of those subject to sterilization laws.

Molly Ladd-Taylor looks at the control of sterilization by local welfare officials in Minnesota during the New Deal. There, sterilization occurred as part of a child welfare policy and was just one part of what its supporters considered a systematic approach to management of social conditions. Procedures for commitment to state institutions as well as the process of public guardianship were closely linked to decisions about which people would be sterilized. In a system that often appeared coercive, patient acquiescence to sterilization could sometimes represent a truly voluntary acceptance of permanent birth control. Ladd-Taylor also points out the distance between the wishes of fervid "eugenicists" in Minnesota and their less ideologically driven counterparts who controlled the sterilization bureaucracy in that state.

Johanna Schoen frames her discussion of North Carolina sterilization history within the story of an operation that occurred in 1968, at a time when most states had all but abandoned the practice. At that late date, the emphasis in North Carolina was on young black women, and those who endured surgery were often known to be victims of rape and incest. Public rhetoric about fears of burgeoning welfare rolls sounded alongside concern about sexual activity outside of marriage. The number of surgeries done in North Carolina was partially explained by a unique feature in that state's law. It was the only state in the country that allowed social workers to petition for sterilization of people living outside of hospitals and asylums in the community at large.

The final essay in this group chronicles the legal case that finally interrupted the widespread use of sterilization in public institutions in the mid-1970s. Gregory Dorr draws our attention to Alabama, another southern state, but one where most sterilization occurred outside the limiting rubric of any "eugenics" law. Dorr links the case of *Relf v. Weinberger* to policies launched as welfare reform during the presidency of Richard Nixon. Illegitimate mothers provided a political target for Nixon just as they had for politicians in Carrie Buck's day, and extending sterilization beyond institutions was accomplished via Nixonian welfare policy, just as it was by the expansive North Carolina law portrayed by Johanna Schoen. Dorr completes his account of the national sterilization politics during the Watergate era with a discussion of the significance of sterilization regulation to reproductive autonomy more generally.

PART 4. EUGENICS IN THE HUMAN GENOME ERA

The book is rounded out with two essays by scholars from the biomedical sciences and law: reflections by a geneticist and a physician/geneticist on the reappearance of eugenic thinking during the human genome era, and a legal scholar's analysis of the ways that eugenic motives are embedded in modern law.

Linda and Edward McCabe approach their topic as scientists and clinicians who are surveying the landscape in their own field of genetics. They are aware that at its inception genetic science was closely aligned with eugenics, and they ask how much our current attitudes and practices mirror those of their predecessors, who embraced toxic features of hereditarian thinking. The McCabes find disquieting similarities in the language of genetic determinism that has reappeared in the age of the Human Genome Project, and they decry both the oversimplified portrayal of hereditary mechanisms and the exaggerated claims made on behalf of "genomics" that have resulted. Government agencies or other entities (like insurance companies) raise particular concerns when they are positioned to exert social or legal control to intrude on the reproductive decisions of individuals.

The final essay, by Max Mehlman, subjects practices he describes as "neoeugenic" to legal analysis. They involve situations when government, through law or policy, encourages, discourages, prohibits, or requires reproduction. Mehlman notes that governments are rarely neutral about reproduction, and in this context active regulation in the form of laws passed (e.g., sterilization laws), court cases decided (e.g., wrongful birth claims), and programs funded (e.g., welfare payments for contraception or tax credits for childbearing) must be considered along with inaction—such as allowing private entities like internet dating services or in vitro fertilization clinics to screen customers based on race, social class, aptitude, or disability. The usual subjects appear as the objects of regulation, and as always, they include the poor, minorities, criminals, and the disabled, as well as those whose sexual practices deviate from common norms.

In the attempt to accurately characterize the ideas and practices that have come to be described as "eugenics," this collection supplies us with much to consider. Four points of particular importance emerge.

First, in practice—as it played out in law and social policy—the definition of eugenics was always changing. To some it meant cutting the tax burden generated by welfare dependent mothers by preventing the birth of more children of poverty. To others, it meant encouraging the most prosperous and successful to multiply, while impeding the replication of the deviant, the disabled, the diseased, or the criminal. Still others used eugenics as a touchstone for their fears that "inferior" racial groups were growing and must be interrupted lest they overrun a less fertile but "superior" race.

Second, making "eugenic" ideas into law was always the result of individual efforts and idiosyncratic motives; the administration of those laws over a span of many years was undertaken by different people with changing motivations. Any of the meanings of eugenics suggested above might take precedence when laws for sterilization, marriage restriction, or racial separation were written, amended, or renewed.

Third, the rationales that lawmakers and other officials voiced in support of "eugenic" practices were clothed in ideas that had popular resonance. What "eugenics" meant for the seven decades it was imbedded in U.S. law depended on what other ideas were used to justify it. Those ideas could include hereditary degeneracy, social and economic efficiency, neo-Malthusian population policy, or "sins of the father" religious determinism. The popularity of *eugenics* over time as a catchword that described the aspirations and fears of those who used it was bolstered by the vagueness of the term itself. It borrowed meaning from the social and political agendas of the people who found practical uses for it no less than from those who first offered it as an idea.

Finally, eugenics made for strange alliances. As the essays in this volume have shown, it could bring together preachers like Oscar McCulloch with educator/scientists like David Starr Jordan. It could also link activists as seemingly disparate as Margaret Sanger and W.E.B. Du Bois. Those who wish to understand what eugenics meant in the twentieth century, as well as those who shout "eugenics" as a heckler's keynote in today's political debates, would do well to review these essays. In them they would find that context is critical; both the idea of eugenics and the practices it led to had many faces.

NOTES

1. Dr. W. C. Rucker, "More 'Eugenic Laws': Four States Consider Sterilization Legislation and Nine Contemplate Restrictions on Marriage—None of Proposed Laws Satisfactory from Eugenic Viewpoint," *Journal of Heredity* 6 (1915): 219.

2. Francis Galton, *Inquiries into Human Faculty and Its Development* (London: Macmillan, 1883), 17.

3. For useful surveys of scholarship on eugenics, see Philip J. Pauly, "The Eugenics Industry—Growth or Restructuring?" *Journal of the History of Biology* 26 (Spring 1993): 131–45; Frank Dikotter, "Race Culture: Recent Perspectives on the History of Eugenics," *American Historical Review* 103 (April 1998): 467–68.

4. The entire quotation is: "We greatly want a brief word to express the science of improving stock, which is by no means confined to questions of judicious mating, but which, especially in the case of man, takes cognisance of all influences that tend in however remote a degree to give to the more suitable races or strains of blood a better chance of prevailing speedily over the less suitable than they otherwise would have had. The word *eugenics* would sufficiently express the idea." Galton, *Inquiries,* 17n1.

5. Mark Adams, ed., *The Wellborn Science: Eugenics in Germany, France, Brazil, and Russia* (New York: Oxford University Press, 1990), 221.

6. Paul Weindling, "The Survival of Eugenics in 20th-Century Germany," 52 *American Journal of Human Genetics* 52, no. 3 (March 1993): 643–49.

7. The title of Daniel Kevles's now classic work, *In the Name of Eugenics* (Cambridge: Harvard University Press, 1985), captures the diversity of objectives that various people pursued, using eugenics as their rationale.

PART ONE

The Indiana Origins of
Eugenic Sterilization

ONE

The Hoosier Connection: Compulsory Sterilization as Moral Hygiene

ELOF AXEL CARLSON

Two important Indiana intellectuals, Oscar McCulloch (1843–91) and David Starr Jordan (1851–1931), provided the rationale for the state's (and the world's) first compulsory sterilization law in 1907. Their writings on "degenerates" influenced physician Harry Clay Sharp (1871–1940), who put the law into practice. McCulloch was a minister of the Plymouth Congregational Church in Indianapolis, Jordan was president of Indiana University in Bloomington, and Sharp was a prison physician in Jeffersonville. This chapter demonstrates the influence of McCulloch's and Jordan's ideas on the eugenics movement and why these influences moved Sharp to champion the sterilization law.[1]

Although Indiana led the world in implementing eugenic sterilization, the idea of state-mandated eugenics was not native to the "Hoosier" state. European and American physicians had considered similar interventions in the last quarter of the nineteenth century. But it was the convergence of degeneracy theory, the replacement of charity with precepts of Social Darwinism, and the popularization of biological measures of "social worth" that allowed supporters to propose sterilization as a rational and enlightened social policy.[2]

Degeneracy theory found its origins in the urbanization surrounding the Industrial Revolution and the social problems created by the late nineteenth century's boom-bust economy. Accidents, ill health, and the tragedies of life (loss of a spouse, becoming orphaned, abusive home environments) that could lead to unreliability at work or unemployment compounded the structural economic challenges that strained American life.[3]

Even as the economy changed, ideas about the nature, scope, and object of charity—a long-standing feature of America's Christian tradition—shifted across the country. After the 1850s, many philanthropists who had initially viewed the less fortunate as worthy objects of assistance came to understand the poor, diseased, and physically infirm as defective in body or mind, often undeserving of charity.[4] This ideological change led to a cultural divide, giving rise to a need among the advantaged (whether with power, health, or wealth) to control the increasingly scorned disadvantaged.

Finally, the biological interpretation of life—holding that the physiological imperfections that led to degeneracy were themselves marks of an individual's inherent biological inferiority—gained great social currency among educated elites who wielded political power. The theory of French scientist Jean Baptiste Lamarck that characteristics acquired during life would be passed down as a function of heredity, led some scientists to the optimistic view that a good environment had the potential to reverse bad heredity. The Lamarckian view that prevailed into the 1880s was challenged and eventually supplanted by a view of heredity developed by August Weismann in Germany. Weismann's experiments demonstrated that the key to heredity was lodged in the reproductive tissue and that heredity was isolated from most environmental influences that change our body tissues and organs. This more restrained notion of genetic malleability from one generation to the next suggested more pessimistic conclusions about the likelihood of ameliorating the conditions of human society.

These intellectual trends combined with new surgical procedures—specifically, less-invasive sterilization techniques that did not castrate the patient—to bolster the popularity of surgical eugenic intervention as an enlightened public policy. The favored method for sterilization in men was vasectomy, a procedure rarely used before the 1890s.[5] Vasectomy provided the means for breaking the chain of heredity not by changing an individual's genetic endowment but by ensuring that only the most genetically fit would procreate, thereby increasing the quantity and distribution of fit traits in the population. Preventing the unfit from passing on their defective heredity was a kind of eugenic public health program.[6]

McCulloch's role was to popularize the shift from an enabling charity to reproductive isolation of those considered unfit to reproduce. Jordan provided biological legitimacy to plans for isolating the unfit, using evolutionary arguments about degeneracy and fitness. Sharp extended the therapeutic powers of medicine by applying ideas of eugenics to the reproductive system. Degeneracy was recast as a disease, and vasectomy represented an efficient, humane way to prevent its transmission.

THE RISE OF DEGENERACY THEORY

Degeneracy theory arose about 1600 in response to a problem that had been growing in Europe since the Protestant Reformation. Prior to the Reformation, the tradition had been for the church, and not the state, to care for the poor and the unfortunate. England's Queen Elizabeth I secularized public philanthropy by authorizing Poor Laws, which shifted the responsibility for the destitute from the church to individual counties. By the time of the Industrial Revolution some 250 years later, both the numbers of the poor and the taxes to care for them had increased dramatically. Industrialization, urbanization, and the attendant political unrest led to legislative attempts to reform the Poor Laws. By the 1830s, most middle-class taxpayers were persuaded that "degenerates"—paupers, criminals, prostitutes, alcoholics, psychotics, and the mentally retarded, along with people considered physically disabled—were not the result of social or economic problems that could be ameliorated by Poor Laws; rather, they were medical problems. Isolating the unfit or deporting them to colonies was one solution. The New World was the first dumping ground, and Australia was the second, as the movement to isolate the unfit gained approval.

These ideas about fitness gained currency in the United States after the Civil War. During the late nineteenth century, English social theorists advanced many strategies to resolve the problem of the poor, including the notion that the West was a "safety valve" for overcrowded urbanites. Sometimes the poor were auctioned off to do labor for little or no pay for those who won the bid. Sometimes the "unfit" were herded together in poor houses. Almshouses and workhouses were supposed to be self-supporting, but frequently their residents did not generate sufficient income to meet their needs, and the facilities were forced to rely on local or state taxes to cover their operating expenses. Sometimes the unfit were placed in asylums, but that, too, was expensive. Increasing numbers of institutionalized individuals required more generous state support to build and maintain adequate facilities.[7]

Who were these unfit people? Some were children, too young (especially if orphaned) to support themselves. They may have suffered from a chronic illness or birth defects that made their employment difficult and their needs more than a family could provide. The fragile economic balance of the family wage could be shattered when women lost their husbands to accident, illness, or the all-too-common desertion. An adult might be mentally retarded, blind, deaf, or physically or mentally incapacitated. Before there was a science that could discern and interpret the causes of these disorders, most people assumed that misfortune was an act of God. As society secularized and advances in science and medicine offered interpretations that seemed equally plausible or even superior to religious determinism, degeneracy theory emerged, relying on the authority of science to explain the existence and experience of society's less fortunate members.

Instead of asking, "Why does God allow bad things to happen to good people?" American intellectuals increasingly asked, "What is it about an individual's inherent biology that makes him bad?" Was "nature" or "nurture" responsible for misfortune? Bad environments could be reversed, but often when that "social therapy" was done through charity and asylum reform movements, the "treatment" was expensive, and many of those treated did not respond. The blind remained blind. The retarded remained retarded. The psychotic remained psychotic. Paupers lapsed back into their impoverished neighborhoods and degraded, poverty-stricken habits.

By the 1880s, many environmental reforms had been tried and found wanting. The seeming intractability of degeneracy and its social consequences frustrated those looking for a quick response to the problems of modern living. At this point reformers and physicians, newly emboldened by their successes in public sanitation, public hygiene, and public health, stepped in to promote a new, simple, preventive remedy for longstanding social problems. Hygiene became the watchword of efficient reform.[8] Those who were most touched by degeneracy and therefore most in need of reform clustered together in problem families or clans. Oscar McCulloch became known for his discovery of one such clan.

OSCAR MCCULLOCH AND THE CHARITY MOVEMENT IN INDIANAPOLIS

Oscar McCulloch's father wanted him to pursue a business career, but McCulloch felt called to the church. Raised and educated in Illinois, McCulloch's first church was in Wisconsin. Moving to Indianapolis, he took over a church with falling membership, uninspired services, and heavy debts. He used the business skills his father had emphasized to fill the church coffers and to provide many social services for the surrounding community, including a free library, a soup kitchen, housing, jobs, childcare, and a literacy program for the poor. McCulloch's reform activism soon filled the church with parishioners, and as he gained recognition in Indianapolis society, he used his prominence and power to organize the city's scattered, redundant charities and make them more effective. He befriended the leaders of the Knights of Labor and championed the poor laborers who entered the union movement. Meanwhile, he took an interest in science and attracted intellectuals to his church. He was comfortable with the findings of science, including Darwinian evolution, believing that humans had a lot to learn about God's universe and that science was a major source of new insights.

McCulloch was proud of his successes, but in the early 1880s he encountered one failure that puzzled him. He tried to help a "tribe of degenerates"—an impoverished group that lived along the banks of the White River who had first come to Indiana when it was not yet a state. They called themselves the Tribe of Ishmael, and they traveled in caravans during the spring and summer.[9] They wintered in Indianapolis, supporting themselves with odd jobs. They owned no fixed property and shunned permanent jobs. The tribe claimed descent from itinerant English tinkers, escaped indentured servants, escaped or freed slaves mostly of the Fulani migratory African tribes, and Native American Shawnees, another migratory tribe. The Ishmaelites dressed in colorful garments out of keeping with conventional, middle-class sartorial norms, and they loved their nomadic tradition, though it clashed with the stolid, staid lifestyle valued by reformers. McCulloch could not understand why they preferred to migrate rather than settle down. They lived on the margins of society, paradoxically flouting the obvious benefits of joining the middle class through thrift, industry, and sobriety. He felt that their shiftless behavior evinced an essential degeneracy of character and constitution.

McCulloch viewed the Tribe of Ishmael as a parasitic race with a peripatetic lifestyle. The Tribe grew in number because it was supported by the charity of others, rather than its own thrift and industry. The Ishmaelites were inordinately represented on the tax rolls because their chronic penury, exacerbated by winters in economic lean times, left them dependent on public assistance. McCulloch thought the Ishmaelites incorrigibly defective and believed they should be isolated and prevented from reproducing. Taking a page from the Bureau of Indian Affairs, McCulloch even entertained the idea of removing Ishmaelite children from their parents and raising them in "upstanding" foster homes to see if the cycle of nomadic living could be overcome by a proper environment. Ultimately, emphasizing his judgment that the Ishmaelites were subhuman, he compared them to "devil grass" or weeds that could only be controlled by being uprooted. During regular conversations, David Starr Jordan, a biologist who had joined McCulloch's church, confirmed the minister's conclusions.[10]

DAVID STARR JORDAN AND THE EVOLUTIONARY INTERPRETATION OF DEGENERACY

Jordan received his bachelor's degree at Cornell University and a master's degree at Harvard. In 1875, he took a medical degree at the Medical College of Indianapolis, a one-year proprietary program typical of American medical education before its great late nineteenth-century reforms. Jordan was on the faculty of Butler University in Indianapolis when he met McCulloch. During those years, Jordan's educational pedigree and his growing repute in academic circles marked him as a promising intellectual who was beginning to attract attention.

Jordan was both talented and ambitious. He published frequently in his field of ichthyology and explored the evolutionary histories of North and South American fish. He left Indianapolis for Bloomington in 1879, joining the faculty at Indiana University and ascending to the university's presidency in 1885. While in Indianapolis, he regularly encountered McCulloch at the only church he ever attended. Jordan felt close to Mc-Culloch intellectually and corresponded with him until McCulloch's death in 1891. It was McCulloch who first made him aware of the Tribe of Ishmael, and Jordan provided references on parasitism and evolution for McCulloch to use in his quasi-biological interpretation of the tribe as an example of degenerate "human parasitism."

For 30 years—at Indiana University, and later at Stanford—Jordan taught a course that he called "Bionomics," a senior capstone course on the social and ideological impact of biology. In the course, Jordan addressed contemporary issues through applied biology. Influenced by Francis Galton's ideas on human variation, Jordan reminded his students that while they were intellectually superior to most other people, they also were differentiated among themselves—only a few would become eminent, and most would fill the ranks of America's able but not terribly distinguished professionals.[11] Just as some people are born to be physical giants or dwarfs, Jordan argued, some are born to rank as the best or the least mentally.

It is thus no surprise that Jordan chaired the first committee on eugenics for the American Breeders Association in 1909.[12] Moreover, many of his books promoted a eugenic outlook on society. He felt it was an evolutionary obligation for humanity to cull the least productive of its members and to encourage the best and the brightest to reproduce more of their kind. At the same time, however, he was a pacifist; eugenics taught him that countries recruited their ablest young men to go out and kill each other off in war, artificially and disproportionately allowing the dull and physically weak, out of harm's way safe at home, to procreate. Jordan's move from Indiana to the inaugural presidency of Stanford University—founded with the money of a rational, scientific railroad magnate—ensured that he continued to be widely read across the nation as educated elites tracked his efforts at this promising new bastion of modern education.[13] The story of the Ishmaelites, promulgated by McCulloch with assistance from Jordan, continued to resonate in Indiana, the home of Dr. Harry Sharp, even after McCulloch's death and Jordan's western relocation. Sharp would become famous for leading the movement to eradicate such degenerate clans through eugenic sterilization.

HARRY CLAY SHARP AND THE
HYGIENE MOVEMENT

Sharp was born and raised in Indiana, but he completed medical school in Louisville, Kentucky, in 1896. As a student, he learned of the new public hygiene movement initiated by German medical schools, relying to a major degree on the work of bacteriologist Rudolf Virchow. In addition to his cytological discoveries, Virchow had championed the novel idea that it is a government's duty to protect and care for the health of its citizens. He initiated meat inspections, garbage disposal, purification of water supplies, and free health examinations for all schoolchildren in the name of public health and hygiene.

The state of American medical education in the late nineteenth century was generally unsophisticated, and the most promising American physicians did postgraduate work in Europe, especially seeking out appointments in Germany to study with Virchow or his students. Although Sharp never made this intellectual pilgrimage, he did the next best thing: he read and studied the latest publications about preventive medicine in the important and increasingly reliable American medical journals.

Sharp's first job was prison physician in Jeffersonville, across the Ohio River from Louisville. The physical conditions of the prison and its inmates, suffering from endemic disease and malnourishment, appalled Sharp. Many prisoners died of tuberculosis, typhus, and typhoid fever. The kitchen was inadequate to prepare proper meals, and prisoners ate a watery swill better suited for feeding pigs. At night, inmates tossed their excrement into a trough that ran between the cells; in the morning, a guard would hose the urine and feces out of the trough.

Sharp documented the horrid abuses he witnessed in biannual reports to the governor, prepared in the prison print shop. He charged that by neglecting the health of those for whom it cared, the state was guilty of homicide. While prisoners paid their debt to society, they surrendered their freedom but need not surrender their humanity, Sharp said. He argued that the prison's inflated death rate was largely preventable by simple, relatively inexpensive reforms. He asked for sanitary privies to be installed in the cells, the construction of an adequate kitchen, and the provision of a balanced diet. The governor and the legislature acquiesced, and in later years Sharp was pleased to report a dramatic reduction in the death rate.[14]

In addition to practical reform in the prison, Sharp remained interested in the theoretic foundations of mental hygiene. His first scientific publication focused on hysteria—a theory that attributed emotional breakdowns in women to female anatomy. Men were also subject to nervous exhaustion caused by the hectic pace of modern industrial and urban life. Collectively, European physicians referred to these mental symptoms as neurasthenia, a symptom of degeneracy. Many important thinkers, such as English philosopher Herbert Spencer, who coined the term *Social Darwinism,* believed that society should purge itself of degenerate components. One way of accomplishing such a purge for future generations was through the surgical solution of sexual sterilization.

SETTING THE STAGE FOR
COMPULSORY STERILIZATION

In 1899, Sharp read an article in the *Journal of the American Medical Association* by Albert Ochsner (1858–1925), a physician at the Chicago Medical School who had studied with Rudolf Virchow. Ochsner was a founder of the College of Surgeons, and he had been using vasectomies—a procedure developed in Sweden and England in 1890—as an experimental treatment for enlarged prostate glands. Ochsner had several patients who were in their forties or fifties, and they resumed sexual activity after their surgery. This suggested to Ochsner that degeneracy, too, could be treated by vasectomies: severing an otherwise healthy person's spermatic cord resulted only in sterilization, not impotence. Vasectomy offered the possibility of rendering men infertile without subjecting them to the cruelty of destroying their sexual function. Ochsner argued that degeneracy would rapidly disappear if degenerates were sterilized as a routine procedure; accepting the then-current notion that "like begets like," vasectomy would prevent defective men from reproducing "more of their kind."[15]

After reading Ochsner's article, Sharp initiated his own eugenic program by operating on a young man named Clawson. According to Sharp, Clawson's problem was excessive masturbation. Sharp had learned in medical school that masturbation was a cause of degeneracy. That belief was more than two centuries old and held that loss of semen through any sexual excess, including masturbation, would lead to physical and mental degeneracy that would subsequently reappear in the next generation. Less an innovator than an opportunist, Sharp was simply melding what he read by Ochsner with what he had learned in medical school. There was nothing inherently sinister or malevolent in Sharp's approach to medicine or in his concern for his prisoners' health. He reported that Clawson's health improved, he gained weight, he exhibited a more cheerful demeanor, and his excessive masturbation ceased. Sharp concluded that the correlation between surgery and improvement represented a causal relationship: vasectomy had improved Clawson's health and cured his moral degeneracy.[16]

Sharp then prevailed upon Clawson to recruit his fellow prisoners, persuading them to seek vasectomy as treatment for masturbation.[17] Sharp's published reports and papers do not reveal exactly how many vasectomies he carried out before 1907, but it is clear that surgeries began years before the process was sanctioned by law. Beginning in 1901, Sharp had petitioned the governor to urge the legislature to pass a compulsory sterilization law to prevent degenerates from passing on their condition. It took six years, but in 1907 Sharp's pleas caught the ear of Governor J. Frank Hanly (1863–1920), who sympathized with this medical approach to moral problems. Sharp reported 456 surgeries between1899 and 1909; eventually, he claimed to have performed between 500 and 600 operations. Sharp used his success in Jeffersonville to initiate a national campaign for compulsory sterilization. He attended the annual meetings of the American Medical Association and gave talks on the benefits of vasectomies for reducing degeneracy. He no longer stressed vasectomy as a treatment for masturbation, as he had in his first years after reading Ochsner.

Sharp became sterilization's first nationally successful advocate. He was appointed by Jordan's American Breeders Association eugenics committee to study compulsory sterilization. He also developed a close relationship with Harry Laughlin, superintendent of the Eugenics Record Office in Cold Spring Harbor, New York. Through Jordan and Laughlin, Sharp came to know the guiding light of American eugenics, Charles Davenport. Davenport was the founder of the Eugenics Record Office and another charter member of the eugenics committee of the American Breeders Association.[18] But despite Sharp's growing national connections to eugenics, back in Indiana the law passed through his efforts had come under attack.

Governor Thomas Marshall, the Democrat who replaced Hanly, felt the law was unconstitutional. He ordered Sharp to stop performing sterilizations. Eventually, the Indiana courts upheld Marshall's position, overturning the law in 1919. By that time, however, Sharp was overseas in France as a surgeon in the U.S. Army. After the World War I armistice, he became a practicing physician for the Veterans Administration and abandoned his lobbying efforts for sterilization. His last professional post was with the Veterans Hospital in Lyons, New Jersey. As the Nazi sterilization laws became large-scale state policy, Sharp was interviewed about his reactions in a 1937 article in the *Journal of Heredity*. Apparently somewhat embarrassed at the development of sterilization law in the preceding 30 years, he commented, "We did not know enough about science then."[19]

STERILIZATION AFTER THE "INDIANA EXPERIMENT"

Sharp was successful in initiating a movement that eventually led to more than 30 states passing compulsory sterilization laws. Harry Laughlin eventually replaced Sharp as the major advocate for those laws. With the help of legal scholars, Laughlin designed his own Model Sterilization Act, crafted to survive constitutional challenge. The Laughlin model provided a framework for Virginia's 1924 sterilization law, which would be scrutinized by the U.S. Supreme Court in 1927.[20] In the case of *Buck v. Bell*, Justice Holmes spoke for an eight-member majority, concluding that "three generations of imbeciles are enough" and upholding the constitutionality of state-sanctioned compulsory eugenic sterilization. Sharp played no direct role in this drama. By 1927, his role had shifted from eugenics advocate-physician to a practicing physician in a veterans hospital. Although sterilization was about to undergo explosive policy growth, there was uncertainty among the leading lights in mental health and genetics about the relationship among "degeneracy," "feeblemindedness," disease susceptibility, and genetics. Rather than Sharp's keen, bright tool, sterilization was beginning to appear to be a rather blunt instrument. The causes of mental retardation and psychosis did not fit a typical Mendelian pattern; to the degree that hereditary components existed, they were not passed on in an obvious like-for-like transmission. As a result, sterilizing someone who exhibited negative traits would have a limited effect on curbing the incidence of unexpressed, recessive factors carried by healthy people. The economic disaster of the Great Depression multiplied the number of poor, homeless, disabled, and mentally disordered wandering outside of institutions—those who were formerly called "unfit." Moreover, the economic cataclysm laid low many erstwhile "superior" professionals in fields as varied as law, finance, banking, politics, business, and education. In responding to the widespread deprivation, the Roosevelt administration shifted legislators' and the public's views regarding social problems, their origins, and the most efficacious modes of amelioration. Relief and recovery, not prevention and sterilization, became the watchwords of the government's approach to the economic catastrophe.

LOOKING BACK AT THE COMPULSORY STERILIZATION MOVEMENT

It is remarkable that something as malevolent in its outcome as compulsory sterilization was assessed by one of its initiators not in moral terms but as a failure of scientific knowledge. It is clear that at his death Dr. Sharp did not see Ochsner's suggestion or his own advocacy as immoral, although both used or urged involuntary means of carrying out a social experiment for which evidence was inadequate or disputable at the time. Nevertheless, it is too easy to dismiss Oscar McCulloch, David Jordan, and Harry Sharp as misguided zealots, elitists bent on perverting "honest science" into pseudoscience in the name of social control. These men were not monsters, and the parts they played are not roles in a simple morality play where history and circumstance can be portrayed as the clash of pure good and absolute evil. Instead, these men shared a faith in the ability of science to explain the world and a belief that the government should play a greater role in solving social problems by using the tools and explanations that science supplied. They lived in an era before the "rights revolution" of the 1960s and 1970s, at a time when most Americans believed that the good of larger society was greater than the rights of the individual. Sterilization, for these men and many other Americans, was as broadly consonant with the political tenor of their times as it is intensely dissonant with today's political tone. Forgetting this unnecessarily scapegoats these men and misses the real point of investigating their lives: not to place blame and calumny, but instead to understand why solutions that today appear obviously flawed held such immense appeal in the past.

Yet, although one should acknowledge the ways in which Mc-Culloch, Jordan, and Sharp were merely "men of their times," one must admit that they shared a false view of the relationship between society and heredity. The notion that those individuals they described as "unfit to reproduce" would rapidly spread their defects to an innocent population was and is simply false. Ultimately, civilization and human rights were far less imperiled by "the unfit" than by the authoritarian solutions promoted by those who saw in eugenics an anodyne for the problems of modern living. While no one is born with an inerrant Promethean foresight, it is not beyond our capacity to listen to critics, to open legal proposals to vigorous debate, and to go slowly as we adopt scientific approaches to social and moral problems.[21] Better to err on the side of caution and human dignity than to stumble blithely into the path of tragedy and inhumanity.

NOTES

1. The best source on Jordan's life is his autobiography: David Starr Jordan, *The Days of a Man*, 2 vols. (Yonkers on the Hudson, N.Y.: World Book, 1922). For McCulloch, see Genevieve C. Weeks, *Oscar Carlton McCulloch, 1843–1891: Preacher and Practitioner of Applied Christianity* (Indianapolis: Indiana Historical Society, 1976). No biography of Sharp exists beyond William M. Kantor, "Beginnings of Sterilization in America: An Interview with Dr. Harry C. Sharp," *Journal of Heredity* 28 (1937): 374–76.

2. For a comprehensive history of this transition, see Elof Axel Carlson, *The Unfit: A History of a Bad Idea* (Cold Spring Harbor, N.Y.: Cold Spring Harbor Laboratory Press, 2001).

3. The first book-length treatment of degeneracy theory was offered by Benedict A. Morel, *Traité des Dégénérescences de l'Éspèce Humaine* (Paris: Chez J. B. Balliére, 1857). Max Nordau, *Degeneration* (New York: D. Appleton, 1895) was also influential on middle-class thinking in the 1890s. His popularization of neurasthenia as a modern induced disease began in Nordau, *Conventional Lies of Our Civilization* (Chicago: Laird and Lee, 1884).

4. Richard Dugdale, *The Jukes: A Study in Crime, Pauperism, and Disease* (New York: G. P. Putnam's Sons, 1875). While a careful reading of this book shows Dugdale to be an environmentalist, his interpretations were ignored and the work was used to justify hereditarian notions of degeneracy. See Elof A. Carlson, "Richard L. Dugdale and the Jukes Family: A Historical Injustice Corrected," *Bioscience* 30 (1980): 535–39. See also Henry M. Boies, *Prisoners and Paupers* (New York: G. P. Putnam's Sons, 1893). Boies accepted the view that criminals and the poor represented instances of hereditary degeneracy.

5. David Wolfers and Helen Wolfers, *Vasectomy and Vasectomania* (St. Albans, UK: Mayflower Books, 1974). This polemical book gives a good overview of the procedure's development in Europe and then in the United States. The authors err in depicting Dr. Sharp as a somewhat demented and idiosyncratic physician.

6. The surgical procedures for women were salpingectomy and salpingotomy, involving partial excision or removal of the fallopian tubes and subsequent ligation.

7. E. M. Leonard, *The Early History of English Poor Laws* (Cambridge: Cambridge University Press, 1900).

8. Philip R. Reilly, *The Surgical Solution: A History of Involuntary Sterilization in the United States* (Baltimore: Johns Hopkins University Press, 1991).

9. Oscar C. McCulloch, "The Tribe of Ishmael: A Study in Social Degradation," in *Proceedings of the National Conference of Charities and Correction, Held at Buffalo, July 1888* (Indianapolis: Charity Organization Society, Plymouth Church Building, 1888). See also Hugo B. Leaming, "The Ben Ishmael Tribe: A Fugitive Nation of the Old Northwest," in *The Ethnic Frontier*, ed. Melvin G. Holli and Peter d'A. Jones (Grand Rapids, Mich.: Wm. B. Eerdmans, 1977), 97–142. The most thorough explication of the Ishmael story to date is Nathaniel Deutsch, *Inventing America's "Worst" Family: Eugenics, Islam, and the Fall and Rise of the Tribe of Ishmael* (Berkeley: University of California Press, 2009).

10. See David Starr Jordan, *Footnotes to Evolution: A Series of Popular Addresses on the Evolution of Life* (New York: Appleton, 1898), chapters 11 ("Degeneration") and 12 ("Hereditary Inefficiency"). Jordan's ideas about pauperism being a form of human parasitism were reinforced by E. Ray Lancaster, *Degeneration: A Chapter in Darwinism* (London: Macmillan, 1880).

11. The Herman B Wells Library at Indiana University, Bloomington, has several of these unpublished commencement addresses in its collection.

12. The American Breeders Association was founded in 1903 in response to the rising interest in Mendel's Laws, rediscovered in 1900 by three European investigators: Karl Correns, Hugo DeVries, and Eric von Tschermak. The association established its own journal and in 1909 set up a eugenics committee as a third wing of its interests in heredity—the other two being plant and animal heredity. See *American Breeders Magazine* 1 (1913): 126–29.

13. Jordan expressed his eugenic views in "The Blood of the Nation," *Popular Science Monthly* 59 (1901): 90–100, 129–40; Jordan, *The Human Harvest* (Boston: American Unitarian Association, 1907); Jordan, *Unseen Empire: The Plight of Nations That Do Not Pay Their Debts* (Boston: American Unitarian Association, 1912); and Jordan, *The Heredity of Richard Roe* (Boston: American Unitarian Association, 1913). For his pacifist eugenics, see Jordan and Harvey Ernest Jordan, *War's Aftermath: A Preliminary Study of the Eugenics of War* (Boston: Houghton Mifflin, 1914).

14. Sharp's reports to the governor (1896–1912) are included in the biannual Communication of the State Prison Reformatory at Jeffersonville; copies may be found in the Herman B Wells Library.

15. S. A. J. Ochsner, "Surgical Treatment of Habitual Criminals," *Journal of the American Medical Association* 32 (1899): 867–68.

16. Harry C. Sharp, "Vasectomy as a Means of Preventing Procreation in Defectives," *Journal of the American Medical Association* 53 (1909): 1897–1902.

17. Kantor, "Beginnings of Sterilization in America," 376.

18. On Laughlin and Sharp, see Frances Janet Hassencahl, "Harry M. Laughlin, 'Expert Eugenics Agent' for the House Committee on Immigration and Naturalization, 1921 to 1931" (Ph.D. diss., Case Western Reserve University, 1971).

19. Kantor, "Beginnings of Sterilization in America," 376.

20. Paul A. Lombardo, *Three Generations, No Imbeciles: Eugenics, the Supreme Court, and* Buck v. Bell (Baltimore: Johns Hopkins University Press, 2008).

21. See Elof Axel Carlson, *Times of Triumph, Times of Doubt: Science and the Battle for Public Trust* (Cold Spring Harbor, N.Y.: Cold Spring Harbor Laboratory Press, 2006).

TWO

The Indiana Way of Eugenics: Sterilization Laws, 1907–74

JASON S. LANTZER

The 1907 passage of the world's first eugenic sterilization law was an example of interest group–initiated legislation. However, it was also unlike similar legislation during the Progressive Era, such as the prohibition of alcohol, because it did not originate or grow from a grassroots movement.[1] Rather, the legislation passed because a small group of reformers persuaded politicians of both parties to enact the law. While the people who used scientific, moral, agricultural, and economic arguments to get the law passed are important, they are not the whole story. They then built support for sterilization among the general population after the law passed, helping to keep it on the books for nearly 70 years. Analyzing the political processes that produced laws like this is vital to understanding not only the society within which eugenicists lived but also the type of society they hoped to create.

Although there has been relatively little historical attention to Indiana's eugenic past, the state was on the cutting edge of Progressive reforms, including eugenics. Hoosier reformers—both inside and outside the legislature—possessed the necessary mixture of zeal, belief in scientific progress, activist government, and fear of the future for eugenic sterilization to become a reality. At first, the newly enacted law was controversial, and it lacked widespread support among the public, which doubted the well-born science's ability to remake society. After the courts ruled the initial law unconstitutional, its replacement, which integrated relevant scientific advances, enjoyed wider acceptance. With the passage of time and quiet enforcement, the sterilization law no longer attracted political attention. It benefited from the power of inertia; opponents considered the issue "settled." A new round of potentially fractious legislative agitation would be required to repeal the rarely used law; hence a bad idea remained on the books for nearly 70 years.[2]

CREATING THE WORLD'S
FIRST STERILIZATION LAW

Eugenics, the "science of the well born," embodied a fundamental clash between scientific understanding and the Horatio Alger mythology of the American dream. Americans of the early twentieth century believed in the pioneer spirit of the country. Many believed, and more were bombarded with the message, that anyone could make something of himself, no matter what his origins might be. And yet, amidst this celebration of individual initiative and "luck and pluck," eugenics became widely accepted, despite being based on the argument that genes and heredity—not initiative, environment, and effort—determined a person's destiny.[3]

The conflict between the American dream and the eugenic dream created tension over sterilization and remained a part of the unfolding story of Indiana eugenics until the law was finally repealed. The politicians who crafted the first eugenic sterilization law understood this cultural tension, for they lived in a country that was being transformed by industrialization, immigration, and urbanization.

Reformers who sought to mediate the problems of modern society by invoking scientific authority and incorporating it into public policy and law were known as Progressives.[4] Eugenic sterilization, though it seems retrograde and authoritarian today, was a part of the Progressive ethos, advocated by scientific professionals and supported in many states by a broad upper- and middle-class constituency of educated reformers— people who believed that they knew what was best for society. American eugenicists took pride in the fact that they followed "cutting-edge" scientific social theory. As one supporter put it, "Our country has been the pioneer in this movement and is today the foremost champion and advocate of the cause in the world."[5]

Eugenicists believed that the number of people from inferior genetic stock with primitive moral sensibilities was rapidly increasing. The "feebleminded," "morons," and "degenerates" comprised the "defective, delinquent, and dependent classes." They wasted both private and public philanthropy and, some thought, threatened to destroy American civilization itself. Many eugenicists argued that since people could not escape their eugenic destiny, the only solution was to stop the "unfit" from procreating. Eugenic reformers made common cause with public health advocates on the basis of seeking "to strengthen family and civilization by regulating fertility" and promoting the birth of people biologically predisposed to physical and mental health. The question facing eugenic reformers was whether or not society had the will to obey the iron dictates of heredity and do what was necessary to stop the multiplication of the unfit.[6]

The support of many in the scientific community not only gave the champions of eugenic sterilization an aura of prestige but also conferred the power of moral authority. Eugenics became a secular faith among the scientifically inclined who asserted that by obeying the commandments of biology, humanity could usher in a social millennium. Drawing on a quasi-Christian idiom of human perfectibility and salvation, the eugenics message also struck a chord with many religious people, including many liberal, mainline Protestants. In Indiana, where mainline Protestantism (though not always liberal theology) was dominant, the blending of science and religion by eugenicists facilitated eventual cultural acceptance. In fact, the man who helped bring eugenics to Indiana was a minister. The Reverend Oscar C. McCulloch was the senior pastor of Indianapolis's Plymouth Congregational Church. His duties brought him into contact with a group of families he labeled the "Tribe of Ishmael," whom he believed were at the root of the city's criminal activity and social problems. McCulloch concluded from his research on the Ishmaelites that some harsh remedies were in order.[7]

McCulloch's work stimulated elite support for reform. He counted among his friends, admirers, and parishioners some of the most influential people in the state. Among them were David Starr Jordan, the noted biologist and president of Indiana University.[8] Other allies included Governor (and later Vice President) Thomas Hendricks, Congressman William H. English, and Senator (and later President) Benjamin Harrison. Together, these men represented an early bipartisan consensus that would continue for decades and contribute to the support that eugenic measures received at the statehouse.[9]

Eugenicists referred to McCulloch's study to argue for tightening marriage laws. In Indiana, this move coincided with the need to streamline and standardize the marriage certificate process across the state's 92 counties, which was supported by a broad coalition. In 1905 Representative Jackson Boyd of Putnam County introduced Indiana House Bill 118: "A bill for an act regulating the issuance of licenses to marry." The Committee on Rights and Privileges attached amendments to the bill, including one that refused a license to anyone who had been declared an "imbecile, epileptic, of unsound mind or under guardianship as a person of unsound mind." It also applied to "any male person who is or has been within five years an inmate of any county asylum or home for indigent persons," as well as to anyone with "an incurable or transmissible disease." Also barred from matrimony were intoxicated applicants or those addicted to drugs. After only minor discussion, the House voted 52 to 35 to pass the bill.[10] Deliberations in the Senate were similarly brief, and the bill was signed into law with little fanfare.[11] This first victory for "eugenic marriage" set a pattern for future votes on eugenics in Indiana.

Many other states passed tougher marriage laws in the 1910s, but advocates soon realized that such statutes were not enough to halt the multiplication of the feebleminded.[12] They needed another type of professional to advance their cause, one cloaked in the dispassionate rationality of scientific professionalism.[13] As Albert Edward Wiggam put it, "If His [God's] will is ever to be done on earth as it is in Heaven, it will have to be done through the instrumentalities of science."[14]

In Indiana, the "instrumentality" was a cadre of physicians. The group included Brookville's Henri G. Bogart, who promoted sterilization across the state, and Harry C. Sharp, the head physician at the Jeffersonville Reformatory, who popularized the vasectomy. Other important advocates were Dr. Amos Butler of the Board of State Charities, John N. Hurty of the State Board of Health, and Sharp's superior, William H. Whittaker, superintendent of the Jeffersonville Reformatory.[15]

Sharp first performed vasectomies on prisoners for therapeutic, not eugenic, effects. He intended to curb sexual activity, specifically masturbation, which scientists linked to general degeneracy and insanity.

In 1907, the reform cadre decided to validate Sharp's experimental vasectomy practice, but relied on eugenic arguments to drive the legislative agenda. Both Dr. Sharp and Whittaker, who had supported Sharp's early surgeries, lobbied legislators. Sharp told lawmakers that his patients thus far had volunteered for the procedure and that his operations were quick and, though done without anesthesia, relatively pain-free. Whittaker argued that the procedure would prevent crime, cut institutional costs, and free the state from a considerable economic burden. The doctors' advocacy and the lawmakers' knowledge that they had to validate an ongoing practice or face potential liability as well as scandal soon produced results.[16]

Supporters of the sterilization bill identified a fellow medical professional to carry the legislation. Dr. Horace G. Read, the representative from Hamilton and Tipton counties, had been briefed by Whittaker about the Jeffersonville experience and was convinced of the need for a sterilization bill. Perhaps more important, according to one Indianapolis paper, "it is understood that Governor Hanly is in sympathy with the measure and will sign the bill."[17]

Indiana sterilization supporters were aided in their legislative efforts by earlier attempts to pass sterilization laws in other states. Both Michigan (1897) and Pennsylvania (in 1901 and 1905) had attempted and ultimately failed to enact some form of sterilization statute. In Pennsylvania's case, the law had been vetoed despite the support it had in the medical and scientific communities.[18] Indiana sterilization boosters took note. They lined up both legislative and gubernatorial support before they submitted a virtual copy of the bill that failed in Pennsylvania in 1907.[19]

Even with this planning, Read's bill did not have an easy time in the state legislature. It was first sent to the Committee on Medicine, Health, and Vital Statistics. When opposition formed, Read withdrew it and had it sent to his Committee on Benevolent and Scientific Institutions instead. In the end, the House voted in favor of the bill by a margin of 59 to 22.[20] The bill also faced some initial Senate opposition, and in the vote there, the margin was closer still, with 28 senators supporting the legislation while 16 objected.[21] Governor J. Frank Hanly's signature made the bill the world's first eugenic sterilization law. All these legislative machinations provided eugenic reform with a democratic veneer, although the law passed with virtually no public input and relatively little public discussion.

CHALLENGES TO THE NEW ORDER

The Indiana law became the benchmark for the rest of the nation. Supporters believed it would allow the state to resolve its "degenerate problem" and become a national model of social reform. Sharp even wrote an instructional pamphlet on vasectomy that he distributed around the country as he gave public talks on the need for sterilization. Sharp's expert opinion was sought for testimony in legal cases about sterilization. "The Indiana Plan, as it was sometimes called, was appealing because it seemed rational and required no more than an office visit to carry out the operation."[22]

The eugenics cadre worked hard to safeguard their pet reform. To deflect attention from Sharp's early extralegal surgeries, W. H. Whittaker promised Governor Hanley that only inmates sentenced to Jeffersonville after the law's passage would be candidates for sterilization. Whittaker also pledged to use surgery in a "conservative" manner, by establishing procedures that would ensure that only those prisoners who would actually benefit from the operation would be considered. Whittaker assured Hanly that public sentiment favored eugenic sterilization, and he cited letters he had received.[23] But despite these efforts, the law remained vulnerable to changes in the political climate.

In 1908, Democrats won control of the statehouse. The new governor was Thomas Marshall, a lawyer from Columbia City, who soon expressed doubts about the law's constitutionality. Sharp struggled to protect the law as it faced attack. He crafted guidelines to aid his successors at Jeffersonville to continue the sterilization program.[24] Sharp assured the governor that the law was safe from challenge and that none of the men he had operated on would sue since there was "no damage done."[25]

But Marshall also heard from sterilization opponents. One letter he received from a Plymouth, Indiana, lumberyard owner asked Marshall to "stop the brutal mutilation of the inmates of the Indiana State institutions." Noting the powers of institutional officials, the lumberman claimed: "It is an easy matter to get the consent of a man when you have him in your power." Setting aside Sharp's entreaties, Marshall ordered a halt to sterilizations shortly after taking office, and the moratorium lasted for over a decade.[26]

More than the governor's misgivings about the law were at work in halting Indiana sterilizations. Nationally, eugenic sterilization advocates realized that the number of operations would need to increase dramatically for laws to have the predicted societal impact.[27] This would require expanding the scope of the law to increase the number of people liable to be sterilized, which would require building support for enhanced legislation at the grassroots level. Additionally, more information must be gathered on the "feebleminded problem" in the state.

Several states had mounted eugenic surveys meant to assess states' "mental hygiene" and identify the "feebleminded" in the 1910s. Democratic governor Samuel Ralston followed this trend in 1915, establishing the Committee on Mental Defectives (CMD) to study degeneracy in Indiana. Composed of lawmakers and doctors, the CMD confirmed the position of the medical community regarding the wisdom of instituting state eugenic policies. With the aid of New York's Eugenics Record Office (ERO) and the U.S. Public Health Service, the committee eventually classified over 56,000 Hoosiers as prospective candidates for sterilization because of their feeblemindedness, insanity, or epilepsy.[28] Repeatedly invoking the "menace of the feebleminded," the CMD's surveys described the state's problems in such dramatic terms that the otherwise radical nature of eugenic policies was submerged beneath a rhetoric of fear. The committee believed that increased state control of relief efforts was needed to curb local charitable policies that provided an incentive for multiplication of the "degenerate" classes. Although short-term tax increases might be needed to fund the state bureaucracy that would manage the feebleminded, the committee claimed that the costs of eliminating them in future generations would eventually yield savings.[29]

The CMD published reports in 1918 and 1919, but before its final 1922 report was completed, sterilization seemed to be in jeopardy. At the behest of Republican governor James Goodrich, the law's constitutionality finally faced a formal challenge. In 1921 Justice Howard Townsend, a Goodrich appointee, led the Indiana Supreme Court in ruling the 1907 law unconstitutional. The Court decided that denying a public hearing to inmates who faced sterilization violated the Constitution's due process clause.[30] Despite these policy setbacks, voices arguing for eugenic solutions continued to be heard. After a decade of popular discussion of eugenics in the press, in schools, and in churches, most of America was familiar with the language of eugenics. The outbreak of the First World War and concern about the mental fitness of draftees also prompted increased discussion on the topic.[31] The CMD was only a local manifestation of this larger national trend.

Although Indiana's highest court had invalidated the 1907 law, later in the 1920s sentiment in favor of sterilization again ripened. Senator C. Oliver Holmes of Gary, a member of the CMD, became sterilization's new legislative champion. His hometown newspaper, the *Gary Post-Tribune*, advocated a return to sterilization. The paper pointed to the apparently increasing number of feebleminded people in the state and the taxpayer expense that their care represented, arguing that mass sterilization represented the state's only hope.[32]

In 1925, Holmes authored Senate Bill 86, proposing reinstatement of the sterilization policy and creation of the post of "state eugenicist" to oversee its widespread enforcement. The proposal encountered problems shortly after the bill was introduced. Alarmed, Holmes offered amendments that he hoped would mollify the opposition. The Senate voted 31 to 12 in favor of the bill, but the House tabled it.[33] In 1927, Holmes tried again. While the 1925 bill would have granted the state's eugenics officer the authority to sterilize virtually anyone, the 1927 bill authorized only the sterilization of people in institutions and granted them the right to a review process. Holmes found willing sponsors for this less expansive legislation; the Senate passed the bill "with little objection," 35 to 10. The House followed the Senate's lead, passing the bill by a margin of 78 to 8.[34]

The restoration of sterilization in Indiana coincided with a larger national push for new laws that followed *Buck v. Bell*, one of the most infamous United States Supreme Court decisions. The majority decision, written by Justice Oliver Wendell Holmes Jr., placed sterilization of the feebleminded squarely under the police power of the government's public health mandate. The procedure would prevent successive generations of "imbeciles" from burdening the state and society.[35] But Indiana's law was not an echo of *Buck*'s success; the Supreme Court ruling in *Buck* occurred after Indiana updated its eugenics law. What happened in Indiana resulted from success in popularizing and normalizing eugenics.[36]

The eugenic message adapted to changing social and political context. Eugenic reforms had overcome initial opposition by blending moral, economic, and public health arguments together in the 1920s. Eugenics thrived during the Depression by changing its rhetoric from a "top-down" defense of civilization by elites to a "bottom-up" call for personal sacrifice on behalf of the greater good. Indeed, the 1930s saw three times the number of sterilizations than did the 1920s. The fear that "degenerates" might overwhelm the nation remained real for many Americans.[37]

MAKING THE LAW FIT THE TIME

The 1927 sterilization law was not Indiana's final act of eugenic legislation. Despite general acceptance of eugenic tenets, state-sponsored sterilization needed consistent revision in the years ahead. In the 1930s, Indiana's Democratic governor Paul V. McNutt was urged by advocates to create "a sterilization law with teeth." In response to similar entreaties, in the span of six years the law was updated three times, in 1931, 1935, and 1937. The House actually passed two different sterilization bills in 1931: one dealt with the procedure for institutionalizing sterilization candidates; the other authorized the "sterilization of all persons convicted of committing a felony in which human life has been threatened." The Senate endorsed the first by a margin of 31 to 8 but balked at supporting the latter measure.[38]

In 1935, the House considered a bill crafted by a coalition of urban and rural representatives that would permit the "sterilization of persons committed to insane hospitals." Language borrowed from Harry Laughlin's 1914 Model Sterilization Act and incorporated into the Virginia statute upheld in the *Buck* case was repeated in the bill, which prescribed eugenic sterilization to prevent reproduction among "probable potential parents of mentally incompetent or socially inadequate children." A court-appointed physician would "certify as to whether he believes the best interest of society and the individual will be served by sexual sterilization." While the bill created the right to appeal, it also immunized officials and doctors involved from being sued for participating in the process. The bill was sent to the Committee on State Medicine and Public Health, chaired by a physician who was also one of the sponsors, Horace R. Willan, of Martinsville in Morgan County. The committee favorably reported the bill four days later. On March 5, the House passed the bill 61 to 22, with 30 to 9 Senate ratification quickly following.[39] The sterilization bill's easy passage in the upper house can perhaps be explained by the fate of Senate Bill 304, which had called for "the castration of persons convicted of certain crimes." Though this bill had initial support, the controversial nature of asexualizing criminals engendered lengthy debate and narrow defeat (21–23). The House measure, which eliminated procreative capacity while leaving sexual function intact, seemed much more moderate.[40]

In 1937, Hoosier politicians passed legislation that added epilepsy and incurable "primary and secondary" types of feeblemindedness to the list of conditions justifying sterilization. The House passed the two bills 68–15 and 69–12, respectively; the Senate followed suit by margins of 32–8 and 26–12.[41] Only in the 1933 and 1939 legislative sessions did eugenic reformers fail to amend their laws. In 1933, the House considered two bills that dealt with sterilization. House Bill 265 called for the sterilization of criminals in Indiana's prison system. Although the Committee on State Medicine and Public Health recommended the bill's passage, it was eventually withdrawn.

In some ways, the Second World War was a reckoning for the American eugenics movement as it came face to face with the horrors of the Nazi final solution. But in other ways, the 1940s remained an active period for sterilization advocates in both Indiana and the nation. Indiana's sterilization rate hit its highpoints during and just after the war, recording 132, 123, and 159 procedures in 1940, 1945, and 1946. The three-year total of 414 equaled 26 percent of all people sterilized in Indiana between 1936 and 1962. The war did have an indirect impact: as doctors entered the armed forces, it was often difficult to find surgeons to perform the procedures—perhaps explaining the dip between 1940 and 1945.[42]

Sterilization of criminals suffered a setback in *Skinner v. Oklahoma* (1942),[43] when the United States Supreme Court struck down a law to sterilize three-time offenders on "equal protection" grounds. The flaw in Oklahoma's law was that white-collar felons were spared surgery, whereas it was prescribed for petty criminals such as chicken thieves. The Court doubted the value of eugenics in fighting crime, though its skepticism was not extended to the sterilization of the "mentally defective."[44]

In the post-*Skinner* environment, the Indiana Senate attempted to cede all "powers and duties in connection with sexual sterilization" to the Indiana Council for Mental Health. While the legislation cleared the

Senate 43–0, the House sent the bill to its Committee on Social Security, where it died from inattention.[45]

In 1951, the legislature adopted new procedures for appealing a sterilization order. Representatives George S. Diener of Marion County and Betty Malinka of Lake County introduced House Bill 275 to expand the appeals process to include both the local circuit court and the Indiana Council for Mental Health. The bill was approved by the House by a vote of 78 to 10. The Senate also passed it, 32 to 15. On March 5, Democratic governor Henry F. Schricker signed the bill into law.[46] It was the final amendment of eugenic sterilization procedures in Indiana, and it gave more rather than fewer rights to patients. By then it was clear that Indiana lawmakers had gone as far as they were going to go in expanding the reach of sterilization law. When the House attempted to enlarge sterilization's scope in 1955 by extending sterilization to patients in state psychiatric hospitals, the measure died in committee.[47]

REPEAL

The law was now functioning primarily under the control of medical professionals, a policy that had been urged when Sharp and his cadre first pushed for a sterilization law. Indiana's Mental Health Department, created in 1945, oversaw the state's sterilization program. Largely thought to operate above the political fray, the sterilization board consisted of a circuit court judge, a doctor in general practice, a doctor who specialized in the treatment and care of mental diseases, the chief executive officer of the State Welfare Department, and the chief executive officer of the State Board of Health. Each member of this bureaucracy was appointed by the governor. But like most other states, by the late 1950s, Indiana saw fewer and fewer institutionalized sterilizations. Medical professionals and others were starting to question the value of the law.[48]

In a routine legislative cleanup of the statute books, the state legislature formally repealed the 1907 sterilization law in 1963, over 40 years after it was declared unconstitutional.[49] Opponents of sterilization would wait more than 10 years before the remaining sterilization law was repealed.

In 1972 Speaker of the House Otis Bowen was elected governor of Indiana. Midway through Bowen's first term in 1974, James Drews, the representative from LaPorte, Marshall, and Starke counties, authored a repeal measure, House Bill 1238. It passed 65 to 29. In the Senate, the bill was cosponsored by Senators Walter Helmke of Allen and DeKalb counties and Robert Garton of Bartholomew and Johnson counties, giving it statewide support. It passed by a margin of 45 to 5.[50] Garton recalled the repeal process:

> My primary recollection of House Bill 1238 . . . was my feeling at the time that we had made a lot of progress in treating and caring for individuals who were mentally ill or suffered with epilepsy. I realized the old laws allowed the superintendent of a mental health institution or hospital to sterilize patients who were mentally ill or suffering from epilepsy. Even though the superintendent had to receive approval from the institution's governing board, the patient's legal guardian or next of kin, and a Circuit Court judge, it was still amazing that sterilization could be considered as treatment in the best interests of an individual or society. It was equally surprising that the law was not repealed until 1974. . . . If nothing else, the repeal of Public Law 60-1974 proved that public sentiment and even expert advice are sometimes wrong, resulting in unforeseen consequences.[51]

The repeal legislation was signed on February 13, 1974, by Governor Bowen. It brought an end to state-sponsored eugenic sterilization in Indiana. Bowen, a World War II veteran who went on to become an obstetrician-gynecologist and future secretary of Health and Human Services in the Reagan administration, recalled: "When the legislature became aware of it [forced sterilization/eugenics law] (and me as well) [repeal] just seemed the right thing to do and was done with minimal fanfare."[52] Between 1907 and 1974, however, well over 2,500 people had been legally sterilized in Indiana.[53]

NOTES

1. Jason S. Lantzer, *Prohibition Is Here to Stay* (Notre Dame, Ind.: University of Notre Dame Press, 2009).

2. Alexandra Minna Stern, "We Cannot Make a Silk Purse Out of a Sow's Ear": Eugenics in the Hoosier Heartland," *Indiana Magazine of History* 103 (March 2007): 3–38.

3. Diane B. Paul, *Controlling Human Heredity: 1865 to the Present* (Amherst, N.Y.: Humanity Books, 1995), 11.

4. Robert M. Crunden, *Ministers of Reform: The Progressives' Achievement in American Civilization, 1889–1920* (Urbana: University of Illinois Press, 1984); La Reine Helen Baker, *Race Improvement of Eugenics: A Little Book on a Great Subject* (New York: Dodd, Mead, 1912), 108–11.

5. J. H. Landman, "The Human Sterilization Movement," *Journal of Criminal Law and Criminology* 24 (July/August 1933): 403.

6. William Cecil Dampier-Whetham and Catherine Durning Whetham, *The Family and the Nation: A Study in Natural Inheritance and Social Responsibility* (New York: Longmans, Green, 1909), 2–3, 11, 69, 214–15, 220.

7. Oscar C. McCulloch, "The Tribe of Ishmael: A Study in Social Degradation," *Proceedings of the National Conference of Charities and Corrections, 1888* (Indianapolis: Charity Organization Society, Plymouth Church Building, 1888), 154–59; Oscar C. McCulloch, *The Open Door: Sermons and Prayers by Oscar C. McCulloch, Minister of Plymouth Congregational Church, Indianapolis, Indiana* (Indianapolis: William B. Burford, 1892); Genevieve C. Weeks, *Oscar Carleton McCulloch, 1843–1891: Preacher and Practitioner of Applied Christianity* (Indianapolis: Indiana Historical Society, 1976). In addition to McCulloch's study, follow-up reports on the "tribe" were conducted in 1890 by J. Frank Wright and in 1916 by Arthur Estabrook.

8. Alexandra Minna Stern, *Eugenic Nation: Faults and Frontiers of Better Breeding in Modern America* (Los Angeles: University of California Press, 2005), 22, 84–85, 133.

9. Stephen Ray Hall, "Oscar McCulloch and Indiana Eugenics" (Ph.D. diss., Virginia Commonwealth University, 1993), 41, 121, 245.

10. *Journal of the House of Representatives of the State of Indiana, 64th Session of the General Assembly* (Indianapolis: William B. Burford, 1905), 26–27, 38–39, 49. While Hanly did not mention this bill in his gubernatorial address, he did call for stronger divorce laws, the creation of more hospitals for the insane, and the creation of an epileptic colony.

11. Ibid., 253–54, 891–93, 1457; *Journal of the Senate of the State of Indiana, 64th Session of the General Assembly* (Indianapolis: William B. Burford, 1905), 1512–16, 1722–23.

12. Christine Rosen, *Preaching Eugenics: Religious Leaders and the American Eugenics Movement* (Oxford: Oxford University Press, 2004), 67–69.

13. David J. Bodenhamer and Hon. Randall T. Shepard, eds., *The History of Indiana Law* (Athens: Ohio University Press, 2006), 71.

14. Albert Edward Wiggam, *The New Decalogue of Science* (New York: Garden City, 1925), 19

15. *New York Times,* November 1, 1940; Angela Gugliotta, "'Dr. Sharp with His Little Knife': Therapeutic and Punitive Origins of Eugenic Vasectomy—Indiana, 1892–1921," *Journal of the History of Medicine* 53 (October 1998): 371–406; Hall, "Oscar McCulloch and Indiana Eugenics," 254–57; Dennis L. Durst, "Evangelical Engagements with Eugenics, 1900–1940," *Ethics & Medicine* 18 (Summer 2002), http://www.ethicsandmedicine.com; *Indianapolis Star,* September 26, 1921. Other reformers joined these men, including Elizabeth Grannis of the National Christian League for the Promotion of Purity, an offshoot of the Women's Christian Temperance Union.

16. *Indianapolis Star,* March 7, 1907; *New York Times,* November 1, 1940.

17. Jan Shipps, "J. Frank Hanly: Enigmatic Reformer," in *Gentlemen from Indiana: National Party Candidates, 1836–1940,* ed. Ralph D. Gray (Indianapolis: Indiana Historical Bureau, 1977), 238–68.

18. Mark H. Haller, *Eugenics: Hereditarian Attitudes in American Thought* (New Brunswick, N.J.: Rutgers University Press, 1963), 50.

19. Gugliotta, "'Dr. Sharp with His Little Knife,'" 388.

20. *Journal of the House of Representatives of the State of Indiana, 65th Session of the General Assembly* (Indianapolis: William B. Burford, 1907), 367, 437, 613, 1341–42.

21. *Journal of the Indiana State Senate, 65th Session of the General Assembly* (Indianapolis: William B. Burford, 1907), 1302, 1317, 1723, 2220.

22. *Anderson Herald-Bulletin*, February 26, 1996; Harry C. Sharp, "Vasectomy: A Means of Preventing Defective Procreation" (Jeffersonville, Ind., 1909); *Osborn v. Board of Examiners*, 169 N.Y.S. 638 (1918).

23. W. H. Whittaker to Governor J. Frank Hanly, February 8, 1908, "Temporary File: Governor Marshall Papers in Regards to Sterilization," Indiana State Archives, Indianapolis.

24. Gugliotta, "'Dr. Sharp with His Little Knife,'" 396–400.

25. Sharp to Marshall, May 13, 1909, Temporary File: Governor Marshall Papers in Regards to Sterilization, Indiana State Archives.

26. William O'Keefee to Marshall, January 12, 1909, Temporary File: Governor Marshall Papers in Regards to Sterilization, Indiana State Archives; Jason S. Lantzer and Alexandra M. Stern, "Building a Fit Society: Indiana's Eugenics Crusaders," *Traces of Indiana and Midwestern History* 19 (Winter 2007): 4–11.

27. H. E. Jordan, "Surgical Sex-Sterilization: Its Value as a Eugenic Measure," *American Journal of Clinical Medicine* 20 (1913): 983–87.

28. Lantzer and Stern, "Building a Fit Society"; Robert L. Osgood, "The Menace of the Feebleminded: George Bliss, Amos Butler, and the Indiana Committee on Mental Defectives," *Indiana Magazine of History* 97 (December 2001): 253–77; *Indianapolis Star*, December 7, 1923, May 30, 1928, October 23, 1932, September 27, 1938; *Anderson Herald-Bulletin*, February 26, 1996. Members included Dr. S. E. Smith, Dr. Charles P. Emerson, Dr. George F. Edenharter, and Monsignor Francis H. Gavisk.

29. Lantzer and Stern, "Building a Fit Society"; Meetings, December 17, 1915, March 17, 1916, Minutes Committee on Mental Defectives 1915 folder, Committee on Mental Defectives, Indiana State Archives, Indianapolis; Field Worker Files, Series VII, box 1, Eugenics Record Office Papers, Manuscript Collection 77, American Philosophical Society Library, Philadelphia; F. D. and F. H. Streighthoff, "Indiana: A Social and Economic Survey" (Indianapolis: W. K. Stewart, 1916), 200–203; *Anderson Herald-Bulletin*, February 26, 1996.

30. George A. H. Shideler to Governor James P. Goodrich, September 9, 1919, folder 2, box 159, Indiana Reformatory Correspondence, Documents and Reports, Papers of Governor James Goodrich, Indiana State Archives; *Williams v. Smith*, 190 Ind. 526 (1921).

31. Rosen, *Preaching Eugenics*, 94–96, 111, 149.

32. Meeting, August 8, 1924, Minutes Committee on Mental Defectives 1915 folder, Committee on Mental Defectives, Indiana State Archives, Indianapolis; *Gary Post-Tribune*, February 3, 1925.

33. *Journal of the Indiana State Senate, 74th Session of the General Assembly* (Indianapolis: William B. Burford, 1925), 92, 197–98, 203, 321, 330–31, 394, 537–38; *Indianapolis Star*, February 24, 1927; *Journal of the House of Representatives of the State of Indiana, 74th Session of the General Assembly* (Indianapolis: William B. Burford, 1925), 452–53, 604.

34. *Indianapolis Star*, February 24, 1927; *Journal of the Senate of the State of Indiana, 75th Session of the General Assembly* (Indianapolis: William B. Burford, 1927), 285, 356,

618, 1018; *Journal of the House of Representatives of the State of Indiana, 75th Session of the General Assembly* (Indianapolis: William B. Burford, 1927), 565, 601, 842.

35. *Buck v. Bell,* 274 U.S. 200 (1927); Paul A. Lombardo, *Three Generations, No Imbeciles: Eugenics, the Supreme Court, and* Buck v. Bell (Baltimore: Johns Hopkins University Press, 2008).

36. Daniel J. Kevles, *In the Name of Eugenics: Genetics and the Uses of Human Heredity* (New York: Alfred A. Knopf, 1985), 61–63.

37. Wendy Kline, *Building a Better Race: Gender, Sexuality and Eugenics from the Turn of the Century to the Baby Boom* (Berkeley: University of California Press, 2001), 98–99, 107; *New York Times,* June 21, 1932.

38. *Indianapolis Star,* February 3, 1931; *Journal of the House of Representatives of the State of Indiana during the 77th Session of the General Assembly* (Ft. Wayne, Ind.: Ft. Wayne Printing Co., 1931), 173, 229, 259, 386, 569; *Journal of the Senate of the State of Indiana during the 77th Session of the General Assembly* (Ft. Wayne, Ind.: Ft. Wayne Printing Co., 1931), 477, 695, 717.

39. *Indianapolis Star,* February 23, 1935; *Journal of the House of Representatives of the State of Indiana during the 79th Session of the General Assembly* (Ft. Wayne, Ind.: Ft. Wayne Printing Co., 1935), 694, 770, 1082.

40. *Journal of the Senate of the State of Indiana during the 79th Session of the General Assembly* (Ft. Wayne, Ind.: Ft. Wayne Printing Co., 1935), 702, 931, 1025, 1027, 1059, 1068, 1221, 1302.

41. *New York Times,* March 9, 1937; *Journal of the House of Representatives of the State of Indiana during the 80th Session of the General Assembly* (Ft. Wayne, Ind.: Ft. Wayne Printing Co., 1937), 165, 489–90; *Journal of the Senate of the State of Indiana during the 80th Session of the General Assembly* (Ft. Wayne, Ind.: Ft. Wayne Printing Co., 1937), 1191–92, 1215.

42. "Eugenic Sterilization in Indiana," *Indiana Law Journal* 38 (1962–63): 288; Philip R. Reilly, *The Surgical Solution: A History of Involuntary Sterilization in the United States* (Baltimore: Johns Hopkins University Press, 1991), 129.

43. *Skinner v. Oklahoma,* 316 U.S. 535 (1942).

44. See generally Victoria F. Nourse, *In Reckless Hands:* Skinner v. Oklahoma *and the Near-Triumph of American Eugenics* (New York: Norton, 2008).

45. *Journal of the Senate of the State of Indiana during the 86th Session of the General Assembly* (Indianapolis: Bookwalter, 1949), 247, 413.

46. *Journal of the House of Representatives of the State of Indiana during the 87th Session of the General Assembly* (Indianapolis: Bookwalter, 1951), 222–23, 529, 633–34, 710, 974, 1051; *Journal of the Senate of the State of Indiana during the 87th Session of the General Assembly* (Indianapolis: Bookwalter, 1951), 522, 547, 586–87, 672–73, 747.

47. *Journal of the Senate of the State of Indiana during the 89th Session of the General Assembly* (Indianapolis: Bookwalter, 1955), 833; *Journal of the House of Representatives of the State of Indiana during the 89th Session of the General Assembly* (Indianapolis: Bookwalter, 1955), 328, 910–11.

48. C. O. McCormick, "Is the Indiana 1935 Sterilization of the Insane Act Functioning?" *Journal of the Indiana State Medical Association* 42 (September 1949): 919–20.

49. *Journal of the House of Representatives of the State of Indiana, 93rd Session of the General Assembly* (Indianapolis: Hollenbeck, 1963), 264, 273–74; *Journal of the Senate*

of the State of Indiana, 93rd Session of the General Assembly (Indianapolis: Hollenbeck, 1963), 58, 167, 499.

50. *Journal of the House of Representatives of the State of Indiana during the 98th Session of the General Assembly*, 32, 129, 803; *Journal of the Senate of the State of Indiana during the 98th Session of the General Assembly* 310, 332, 348, 686.

51. Hon. Robert Garton to author, August 14, 2006 (copy in author's possession).

52. Hon. Otis R. Bowen to author, September 28, 2005 (copy in author's possession); Otis R. Bowen, *Doc: Memories from a Life in Public Service* (Bloomington: Indiana University Press, 2000).

53. *Laws of the State of Indiana 1974* (Indianapolis: Central Publishing Co., 1974), 262; *Laws of the State of Indiana 1975* (Indianapolis: Central Publishing Co., 1975), 738; Stern, "We Cannot Make a Silk Purse Out of a Sow's Ear," 3–38, 182–210.

PART TWO

Eugenics and Popular Culture

THREE

From Better Babies to the Bunglers: Eugenics on Tobacco Road

PAUL A. LOMBARDO

EUGENICS AND POPULAR CULTURE

From the first decade of the twentieth century until approximately 1940, *eugenics* was a word that most Americans could expect to encounter regularly. Important citizens made the term respectable, and repetition by schoolteachers, doctors, politicians, and preachers made it an expansive term of reference and eventually a part of popular culture. Mentions of "the well-born science" were common in newspapers and popular magazines, in novels, movies, and plays. Yet there was no universal, uniform definition for *eugenics*; the term encompassed everything from proud pedigrees to healthy births. Over time the invocation of "eugenics" became almost clichéd, as it was employed to signal approval for a wide variety of public policy initiatives. By the time a Chicago politician decided in 1915 to run for alderman as "the eugenic candidate," the term had come to stand for "not just 'good heredity' but goodness itself."[1]

In recent years, historians of American literature and popular culture have identified how pervasive eugenic language and themes became in the first third of the past century.[2] As a result of this scholarship, allusions to eugenics in the novels of F. Scott Fitzgerald or its satiric invocation by Sinclair Lewis or Ernest Hemingway are no longer unusual.[3] While such references have increased, fiction writers are rarely credited with affecting the passage of law. One underexplored link in the history of eugenics concerns just such a relationship between the Georgia sterilization law, passed in 1937, and Georgia native son Erskine Caldwell. Both the personal and literary dimensions of Caldwell's life were touched by ideas we now associate with America's eugenics movement. His work was indirectly responsible for a great deal of public debate that ultimately culminated in his home state's sterilization law, the last such law passed in America.[4]

EUGENICS IN GEORGIA

The political history of the sterilization movement in the South was analyzed in detail by Edward Larson in *Sex, Race, and Science,* a work that placed Georgia in the context of several "Deep South" states that took up eugenics.[5] More recently, scholars such as Karen Keely and Betty Nies have described the intersection of popular culture and eugenics, focusing on themes in the work of novelists like Erskine Caldwell.[6] This essay builds upon the insights of those scholars, linking the popular to the political and showing that while Georgia seemed slow to adopt eugenic legislation, it nevertheless was in the forefront among states where "eugenics" was a familiar topic and a mainstay of popular culture. Voices from the press and the pulpit explored the eugenic value of vital statistics and public health, the proper place of eugenics in the schools, the eugenic importance of "fitter families," and the place of eugenic themes in entertainment. Those voices were heard alongside the more commonly remembered, but later articulated, arguments for marriage restriction laws to protect "racial integrity" or laws to impose sterilization on the "unfit" in the name of eugenics. This survey of eugenics in Georgia culminates in an account of Erskine Caldwell's role in the sterilization debates in the 1930s.

From the first years of the twentieth century, news reports focusing on eugenics in Georgia were extensive. Georgians learned that classes on eugenics would be part of the National Corn Exposition in 1913 and that a "department of eugenics" was a critical need within the planned Child Welfare Exhibit in Atlanta that year.[7] When the 1914 Better Babies show proved a "splendid success," leaders of the Women's Christian Temperance Union saw it as the "first step toward eugenics in Atlanta" and predicted that the capital city would be the "pioneer southern city" to embrace the new science.[8] Not everyone welcomed the "wave of agitation over eugenics" that America was experiencing, and some felt that religion was being displaced by a materialistic fetish.[9]

As support for instruction in eugenics gained momentum, the state commissioner of schools was pressed to make a place within the public school curriculum to study the "science of good birth." He declined, saying that some matters were "too delicate to be handled by teachers" and were best left to parents.[10] Others opposed the idea because they judged teaching about the "laws of inheritance" and "scientific marriages" as difficult as the labors of Hercules. Trying to put such knowledge in the hands of children was like asking them to run with scissors or to play "with something that resembles dynamite."[11]

EUGENIC MARRIAGE: MEDICAL CERTIFICATES

But it was hard to escape popular curiosity about eugenics when even baseball-star-turned-preacher Billy Sunday used it to emphasize intergenerational guilt as part of his "sins of the fathers" evangelism.[12] The language of eugenics boosters, mixing nineteenth-century degeneration theory with the newest in Mendelian heredity and Galtonian biometry, provided a convenient shorthand for attacking drunkenness, sexual excess, and socially problematic behaviors of all kinds.

One legislator declared himself a believer in eugenics after seeing *Damaged Goods,* a film billed as a lesson of "sin's consequences" that played in Atlanta to record crowds.[13] Describing the movie as a "wonderful sermon," Colonel Walter Andrews was converted to the need for "strong laws upon the subject of eugenics." He planned to introduce a bill to require a syphilis test and a medical certificate as a requirement of marriage licensing "for the sake of future generations."[14]

When legislation to require a "bill of clean health" of men before marriage was proposed, it generated great controversy. One lawmaker objected that there was "entirely too much reformatory legislation" being introduced. He resisted making couples dependent on the medical profession and "quack doctors" who had no scruples about issuing bogus medical certificates.[15] When an amendment was offered to extend the scope of required testing to women, the bill died after two days of debate.[16] Similar measures designed to protect "the children of tomorrow" were rejected in 1921, 1923, and 1924, often following attempts to expand the required health tests to women.[17] The Georgia Medical Society offered yet another bill in 1928.[18] But like other attempts to enact marriage laws prohibiting unions between potentially "defective" parents, campaigns for marital "social hygiene" in the name of eugenics never succeeded in Georgia.

One commentator said that the fine "scientific" tone of eugenics ignored human experience. She openly satirized eugenic propaganda, particularly the popular notion of "race suicide," which suggested that Americans of Anglo-Saxon descent were having too few children, and that mating with other ethnic groups would muddy the gene pool of "Old Americans." Her "common sense" argument was: "Throw a pile of mud in a stream and about a mile down the stream you don't find anything but pure water. Say do you know the answer to the eugenic stuff? Good food, clean work, fresh air and chuck the booze."[19] But others found a way to more readily adapt the new hereditary science to bolster traditions stretching back to the state's colonial heritage.

THE ONE DROP RULE: THE EUGENICS OF RACIAL INTEGRITY

To most Georgians living at the turn of the twentieth century, white supremacy was a settled reality. There was, however, always the threat that the "supremacy of the Caucasian" could be undermined in the future by "degrading strains of alien blood." Even though southern states maintained antique laws of racial separation, most having been adopted during the colonial era, the need to be vigilant in forestalling the "unspeakable evils that would follow race amalgamation" was an issue revisited regularly. Some even called for national laws to penalize racial interbreeding or "miscegenation."[20] Comparing interracial relations to polygamy, concerned Georgians argued that just as the Mormons had been forced by federal courts to give up plural marriage, Congress could stop race mixers by passing national legislation.[21]

Members of Georgia's medical profession joined their colleagues elsewhere and argued for racial separation by linking public health and eugenics. An infrastructure for more stringent "racial integrity" enforcement was created with the passage of Georgia's public health and vital statistics laws in 1914. Regularly maintained records of birth, death, and marriage provided the framework for tracking data on race and doubled as a way to emphasize the importance of keeping the "germ-plasm" of different races separate.[22] In 1914, the leaders of Georgia's women's clubs urged their members to support the passage of a public health law that was "social, not self-centered" and benefited from "rational eugenic policies, [and] rational social work directed toward reforms."[23] "Public health" had its own cachet; club leaders made no mention of race.

The value of modern, scientifically derived laws in maintaining the color line became more obvious when states like Virginia amended their older "antimiscegenation" statutes using openly eugenic arguments about the need to keep white bloodlines pure. After national leaders of the eugenics movement, like Harry Laughlin, Lothrop Stoddard, and Madison Grant, helped the founders of the Anglo-Saxon Clubs of America pass Virginia's Racial Integrity Act in 1924, eugenicists there turned to other states.[24] The Virginia governor wrote to every other governor in the country, enclosing a copy of his state's new law.[25] Responses to the letter were forwarded to Dr. Walter Plecker, coauthor of the law and Virginia's state registrar in the Bureau of Vital Statistics. Plecker managed the complex problem of registering citizens at the time of birth, policing registration at marriage and death, and guarding Virginia's racial purity.

Atlanta lawyer James C. Davis was a member of the Georgia General Assembly who wanted a law in his state that would mirror the Virginia legislation. Davis requested information on the Virginia law and in response received 50 copies of Plecker's booklet, *Eugenics and Racial Integrity in Relation to the New Family*. This booklet included a copy of the law itself and laid out the eugenical arguments that provided its foundation. At Plecker's suggestion, Davis contacted antimiscegenation activist John Powell with an invitation to speak before the Georgia House of Representatives.[26] Powell's role as coauthor of the Virginia law had already drawn attention in the Atlanta press,[27] and in accepting Davis's invitation, Powell attributed his initial idea for the Virginia law to a conversation he had with an earlier Georgia governor. Powell was therefore particularly excited about the prospect of coming to Atlanta to speak to Georgia lawmakers.[28] In preparation for Powell's visit, members of the press urged legislators to come up with a "practicable scheme that will cut down the production of African American mongrels." But the difficulty of devising a law "with teeth" that could be used to enforce racial purity standards was not lost on advocates.[29]

The Davis Racial Integrity Bill was introduced in the 1925 Georgia legislative session. Although it passed the House of Representatives, it reached the Georgia Senate too late for serious consideration and failed to pass. In 1927, the next full legislative year, Davis shepherded the bill successfully through the process and eventually had the pleasure of reporting to Powell and Cox that it enjoyed unanimous support in the Senate, attracted only three negative votes in the House, and had been signed into law by the governor.[30] Eugenics lent a scientific gloss to legislation to uphold "racial integrity" in Georgia, but its ultimate effect was merely to support existing social relations and bolster a culture of white supremacy. Proposals to enact sterilization laws for eugenic purposes were more radical. Their supporters eventually overcame opposition by amplifying fears of spreading degeneracy and inflating concerns about the moral, social, and economic threats of crime and mental defect.

THE POLITICS OF SURGERY: CUTTING COSTS AND CUTTING PATIENTS

The first stirrings of interest about a sterilization law for Georgia appeared in the 1912 comments of a physician at the Milledgeville State Asylum speaking on "Sterilization: The Only Logical Means of Retarding the Progress of Insanity and Degeneracy."[31] Within a year, Georgia physicians took up the cry for state-sponsored sterilization. Their concern was focused on increasing numbers of the "criminally insane, idiots, rapists or moral degenerates" lodged in state institutions, and their bill was advanced by a physician/legislator as "a wise and humane" version of "scientific legislation."[32]

Declaring that "heredity governs the development of the human race," Dr. W. L. Champion urged his legislative colleagues to "safeguard the interests of the unborn" and guarantee "the priceless heritage of physical perfection and masterful mind" to future generations. He argued that "confirmed criminals" and rapists should be castrated.[33] But as with earlier attempts at eugenic lawmaking, some criticized mandatory surgery, calling it "premature faddism" and "hair-trigger" experimentation.[34] The bill failed.

Proclaiming the "medical gospel," one doctor declared eugenic sterilization the centerpiece of the movement for public health laws. Holy

Scripture demanded eugenics laws, he said, "so that the creature of the future may be a better specimen of manhood and womanhood; that there may be fewer inebriates and cripples, that our alms houses, hospitals, penitentiaries, chain gangs and asylums may have fewer inmates, and that our streets may be free of beggars and perverts."[35]

Since the next best thing to sterilization was segregation of the "unfit" to contain their fertility, some pressed for more extensive use of intelligence tests in the public schools to identify "feebleminded children" and place them in state institutions.[36] Soon there was lobbying for an institution to house mentally deficient children. Institution building was expensive, but supporters were quick to point out that the cost of housing was less than the cost of caring for criminals who would be the inevitable children of "defectives" living at large. Provision for the feebleminded was sold as a step in the battle against crime.[37]

In a series of essays printed in the *Atlanta Constitution*, the superintendent of Georgia's Gracewood Training School for Mental Defectives reemphasized the need to identify and segregate feebleminded people living in the community. Repeating conventional wisdom among eugenicists, he argued that feebleminded girls were the source of venereal disease and illegitimacy, that more than 40 percent of prostitutes were feebleminded, and that "30 percent of the children in orphanages were defective."[38] The social costs generated by the feebleminded were the lynchpin of his argument for institutionalization. He said that money was wasted on attempting to rehabilitate defectives in reformatories, or trying to educate the defective in schools, yet too little money had been spent to maintain or expand the state facility that would—by quarantining defectives and preventing their mating—represent a step in preventing those costs.[39] Society should look at the "occasional fool" not as a curiosity that each community should tolerate but as an "active, dynamic enemy to the community life, who goes on propagating his kind with terrible fertility, corrupting the law, morals and health of society, never paying his way, and always living at the expense of others."[40] But the ultimate preventative, embodied in a eugenic sterilization law, remained elusive to Georgia's eugenic reformers. By 1929 more than 20 states had adopted sterilization laws; that same year the Georgia State Asylum was carrying a deficit of almost $2.5 million. As the budget shortfall was announced, a sterilization law was proposed.[41] But again, it failed to pass. Three years later, the Georgia Medical Society offered another sterilization bill. It included provisions for appointment of a medical board that would choose candidates for sexual surgery.[42] It also failed.

The year 1932 was the deepest point of economic crisis for most of the United States. Although the cloud of worldwide depression settled over Georgia earlier and lasted longer than it would for the rest of the country, serious consideration of sterilization in Georgia did not gain traction until most other states already had sterilization laws. Perhaps at first Georgia lawmakers were unconvinced by the eugenicists' apocalyptic vision; perhaps voters did not favor drastic medical interventions—even upon the most marginal citizens. But when those convictions finally changed, at least part of the reason for the popular acceptance of sterilization in Georgia was due to the controversy engendered by novelist Erskine Caldwell, a Georgia native.

STERILIZATION LAW IN GEORGIA: FROM TOBACCO ROAD TO THE CAPITOL DOME

Caldwell's father, Ira, was a teacher and a Presbyterian minister who also worked as a journalist. Beginning in 1926, he wrote a weekly column for the *Augusta Chronicle* in which he invited his readers to consider the issues of the day. "Let's Think This Over" became a regular feature of the paper, and as Erskine Caldwell's reputation as a writer matured, the senior Caldwell developed a following of his own, writing about topics as varied as lynching, the chain gang, and high interest rates. He invariably took the position of social progressive, a bold posture from his home in rural Georgia.

In 1929, Dr. John Bell spoke to the national meeting of the American Psychiatric Association in Atlanta. Bell was well known for his victory in the famous U.S. Supreme Court case of *Buck v. Bell,* which had settled the constitutionality of eugenic sterilization laws. Bell's talk, advocating the sterilization of epileptics, was reported in the press.[43] Only a week later, Ira Caldwell raised "The Eugenics Question" in his column. He doubted the value of eugenic sterilization, noting that some state experiments in surgery were launched with the expectation that "the race could be lifted to higher levels of intelligence" by surgery on "the mentally deficient." Caldwell claimed that new research showed many less promising children had been born of prosperous parents, calling into question the surgical solution. Caldwell conceded that it was impossible to separate environmental influences from those of heredity. So he pressed to clean up the environment in which poor children were raised, urging at least as much attention to "bad social inheritance" as to "bad blood." Ira Caldwell's positions on eugenics were nuanced. He did not dismiss eugenics entirely, but was skeptical about its immediate utility. "Probably the eugenic laws will in the end render civilization a service," he said, "but at the present time the science is too new, too uncertain to be wholly reliable."[44]

Caldwell then began an experiment, convinced that social amelioration was the key to attacking poverty. He helped a family of desperately poor farmers move from their hovel on the fringes of the county closer to the village he lived in with all of its trappings of civilization—schools, shops, and churches. He arranged a job for the father, school lessons for the children, and had the family enrolled as members of a local church. The whole community pitched in to better the family's lot, yet the experiment proved a failure. The children dropped out of school, the father quit his job, and before long the family had moved back to its ruined shack.[45]

Caldwell despaired and seemed to have given up on the idea that changing the environment could change the motivation and habits of people he characterized as sunk in the "quicksands of ignorance." He worked out his disappointment by writing a series of articles, five in all, published in *Eugenics: A Journal of Race Betterment*, the official journal of the American Eugenics Society. The series carried the fictitious name Caldwell had assigned to the family, "The Bunglers," and it was illustrated with pictures he had taken of the family.[46]

The articles appeared alongside the journal's other discussions of "dysgenics," describing in detail the distressing conditions in which the Bunglers lived. Not altogether pessimistic, Caldwell suggested that the Bunglers, and millions like them, were created by a complex set of factors including "social, economic, and biologic forces" far beyond their control.[47] Caldwell's series was not a eugenic jeremiad about "poor white trash" and the way in which their plight was determined by the inevitable workings of biology. He claimed instead that despite their many shortcomings, the Bunglers were "honest, hard-working people" who did not drink and rarely got into trouble with the law.[48] Nevertheless, Caldwell ended his comments on a much more negative note. Too many Bunglers were afoot, he said, and they were "to all practical purposes, idiots" who dragged society down. They contributed to the "sociological morass in which the society is more and more engulfed." Perhaps as many as 50 percent of these social "idiots" could be rehabilitated over time. But in the short run, sterilization of them and others of their ilk would lessen the pressure on society to carry the burdens they generated. In a passage consistent with the pronouncements of most supporters of eugenic interventions, Caldwell concluded, "Ignorance, stolid stupidity, thick necks, and low brows are the greatest perils of a republic."[49] Unless these dangers were addressed, the problem would not abate.

Few people would read Ira Caldwell's essay on the grinding poverty of his corner of Georgia; fewer still could understand his struggle with his beliefs. He was pushed alternately toward compassion and a sense of responsibility for his clan of Bunglers, then away from them in disgust and a final admission that only eugenics would offer an appropriate solution to the replication of their kind. In many respects, Caldwell's inner turmoil mirrored that of his region—pulled toward the attractions of modern, industrial life, but repelled by its social leveling and racial intermingling. In contrast, Caldwell's son Erskine would be read by millions; he would reject the eugenics "solutions" he learned both at college in Virginia and as a lament of despair from his father.

Erskine Caldwell was born in Georgia, and he grew up on the state's eastern flank in the village of Wrens, 30 miles from Augusta and the South Carolina border. There he saw rural life as it was lived, riding at times with his father through the countryside, ministering to landowners and sharecroppers alike.

Caldwell's grandmother's lineage qualified him for a scholarship funded by the United Daughters of the Confederacy, which paid his way to attend the University of Virginia. There he was exposed to some of the giants of the field of eugenics, including Robert Bennett Bean.[50] Caldwell visited the asylums and poorhouses in the countryside around "Mr. Jefferson's University" with Bean and saw firsthand how the new science of eugenics worked. He also learned how the problematic "germ-plasm" of defective citizens contributed to their station in life.[51]

No doubt as a result of his father's influences, Caldwell was a race liberal, something that could not endear him to the Anglo-Saxon Club sympathies of the Virginia campus in the 1920s. As someone whose own limited resources landed him on one of the lower rungs of the social ladder at the patrician university, he was forced to work as a janitor in a pool hall to make ends meet. Taking a job that generally fell only to black servants hardly increased his standing in the hierarchical world of Virginia privilege, and Caldwell's first year at the university could not have been pleasant.

Caldwell developed his skill for blunt social critique early in his writing career. In his first published article, written in college at Virginia, Caldwell berated his home state, describing life there in a piece entitled "Georgia Cracker." He railed about Georgia as a place "whose inhabitants do cruel and uncivilized things; whose land is overrun with bogus religionists, boosters, and demagogues; whose politics are in the hands of Klan-spirited Baptists; and yet whose largest city boasts of being 'the greatest city in the greatest state in the world.'"[52] But for the fact that it was printed in a tiny, obscure journal, Caldwell's comments would certainly have drawn the wrath of fellow Georgians.

Caldwell dropped in and out of college at Virginia, finally leaving two years short of graduation. Then he worked in Atlanta as a reporter at the *Atlanta Journal,* learning how to write alongside Margaret Mitchell, later the famed author of *Gone with the Wind.* In 1931 Caldwell left Georgia and moved to New York City, determined to write about the vision that fired his father's reformist passions and ultimately his frustration. Erskine had seen with his own eyes "the poverty and hopelessness and degradation" in the rural South.[53] He left the familiar settings of his youth for the gritty realities of New York, taking some of his own experience and sense of the gritty realities of the rural South with him.

The story of the Bunglers that Ira Caldwell sketched in *Eugenics* became the basis for Erskine Caldwell's novel *Tobacco Road.* Jeeter and Dude Bungler found a new, fictional life as Jeeter and Dude Lester. Other characters from the eugenics journal reappeared in the novel with barely masked physical marks of defect and behaviors that mirrored the real people Ira Caldwell had described. The novelist breathed dramatic life into the people of his home county, picturing them in rundown hovels reeking of squalid poverty and adding touches of sex and scandal guaranteed to shock 1930s sensibilities. Caldwell became famous when *Tobacco Road* was published. He would later be criticized in the South, particularly in his home state of Georgia, for the unflattering, scandalous, pulp fiction vision of the region he created.[54]

THE TOBACCO ROAD TO
GEORGIA STERILIZATION

About the time the furor over *Tobacco Road* was settling, an article appeared below a banner headline in the Atlanta Sunday paper. It explained that sterilization would be one of the major legislative objectives of the Medical Association of Georgia for the year, and the program was described by the Association's president, Dr. Charles H. Richardson. Richardson repeated the claim that institutional care for social defectives was too expensive. The proposed law could eventually "save the government a billion dollars a year." Doctors who supported the proposal wanted a commission of their colleagues to decide who would be sterilized. "I would not be in favor of sterilization should it be within the power of politicians to control any part of it," said Dr. W. E. Barber, a former president of the Fulton County Medical Society. Finding a German example he hoped to emulate, another medical society spokesman proclaimed that the "sterilization project of Hitler in Germany is a step in the right direction." While the recent Nazi government enactment "might seem a bit drastic on the surface," he said, "it is being used wisely," and it actually was less expansive than sterilization laws in some American states.[55]

Doctors who initiated the sterilization campaign conceded that the lay public might be able to do a better job of swaying legislators than the professional men had.[56] To that end, Emory University biologist Robert C. Rhodes lectured Rotarians and Masons about eugenics, calling sterilization a "matter of good citizenship."[57] Momentum built for the measure as members of the Georgia Humane Society also pressed for a state sterilization law.[58] These pleas gained added force as the state cut back on funds for institutionalization. E. E. Lindsey, chairman of the State Board of Control for Charitable Institutions, announced that because of reduced state revenue, there would be a 25 percent reduction in the budget appropriation for his agency. That cut came despite the fact that "mental deficiency appear[s] to be on the increase." Lindsey identified sterilization as the best "means of checking the rapid increase in insanity." In Georgia, like other states, sterilization was seen as a budget management, cost-cutting measure.[59] Lindsey argued that sterilizing institutional patients would help "to reduce the burden of taxation."[60] Driving the point home, Lindsey emphasized that hereditary defects were a "financial problem for the state government and the taxpayers."[61]

While the campaign for sterilization in Georgia grew, Erskine Caldwell was living in New York. Caldwell's novel *Tobacco Road* had been adapted for the stage and had been playing on Broadway for two years when he wrote a newspaper series documenting the poverty he had portrayed in his fiction. The *New York Post* published his four articles, beginning with a scene of a "poverty-swept" Georgia landscape, where "children are seen deformed by nature and malnutrition, women in rags beg for pennies, and men are so hungry that they eat snakes and cow-dung." According to Caldwell, the state stood idly by providing no relief as rural citizens starved. Caldwell characterized Governor Eugene Talmadge as a "dictator" who did not believe in "relief." Caldwell charged that city-dwelling Georgians denied the existence of the naked denizens of the rural areas and their "deformed, starved, and diseased" children.[62]

A subsequent Caldwell article decried the economic exploitation of rural laborers under the oppressive sharecropping system that often resulted in debt peonage.[63] Other Caldwell dispatches portrayed teenage girls dying from anemia and tenant farmers beyond the reach of government relief.[64] Caldwell's final essay was the *coup de grace*. He described a two-room house populated by three families of farm laborers. One small child, afflicted with rickets and anemia, licked an empty paper bag that held only the smell of its previous contents—hog fat. His belly was swollen with malnutrition; "he was starving to death." In the other room, two infants lay on the floor in front of the fire. With no other sustenance, they attempted to nurse from the family dog, repeatedly returning to suck "the dry teats of a mongrel bitch."[65]

Caldwell had weathered earlier criticism from Georgians for his portrayal of life in his native South in novels like *God's Little Acre* and *Tobacco Road*. During occasional visits to family, the local press grudgingly acknowledged him as "a writer of considerable note" who had "carved himself a niche in the literary hall of fame."[66] But when reports of his new crusade as muckraking reporter reached his home state, the locals took great offense. Declaring Caldwell's account "untrue" and "unfounded," they demanded an investigation of conditions in Jefferson County to disprove Caldwell's libelous portrait. Georgia boosters, "incensed" over Caldwell's "sordid tale of squalor and depravity," rose to defend their region's wounded pride. A government relief agency bureaucrat disputed Caldwell's report, saying he was "making money at the expense of his own home people." County commissioners promised a full investigation of all families in need of assistance.[67]

Calling Caldwell's earlier work in *Tobacco Road* "grossly overdrawn," the *Augusta Chronicle* quoted in detail from the *New York Post* series, finding the conditions Caldwell depicted beyond belief. The "outstanding citizens of Jefferson County" would not allow such "wretchedness, poverty, and depravity to exist," the editorial declared. The paper demanded a grand jury investigation to prove that Caldwell was merely a turncoat sensationalist and to absolve the local citizenry of the implication that they were "heartless heathens."[68]

Some Georgians decided to take the battle north, writing directly to the *New York Post*. One man questioned Caldwell's account of men eating cow dung, saying that he had never seen such a thing, but he left open the possibility that the novelist was "referring to his own experience." The defective system of poor relief, he said, was due to the number of "Yankees" who administered the welfare system and, in an attempt to effect racial equality, distributed "much to the Negroes and little to the whites."[69]

But Caldwell's report was not unique. Another series investigating conditions in the South was begun by the Scripps-Howard news chain, and the story of "Bootleg Slavery" was featured in *Time* magazine, complete with excerpts from the Caldwell series. The magazine supported Caldwell's assessment and concluded that "a vast stretch of the South was the scene of humanity hit bottom." Conditions in the region were so bad that the inhabitants could not even get "the three M's—Meal, Molasses and Meat—a diet that nourishes pellagra but not men." Now the battle was joined in earnest.[70]

Georgians attacked the magazine, calling its repetition of the Caldwell slanders "willful exaggeration or inexcusable ignorance," "farfetched tripe," and labeling Caldwell himself a "neurasthenic egomaniac." However, Caldwell's father, Ira, supported him, sending a telegram that asserted: "Erskine Caldwell's story essentially true."[71] A father's defense was expected, and it did little to assuage bruised Georgian pride.

The Augusta newspaper charged ahead with the promised investigation. To its surprise, it found that some families were in utter need of rehabilitation. Some, "living in want and squalor, [were] victims of their own shiftlessness and ignorance." Ira Caldwell guided reporters to some of the farms Erskine had described, where they confirmed that some homes were "unfit for human habitation." They also uncovered "unmistakable evidences of malnutrition, disease and moral degeneration." "Several prominent citizens" agreed on "corrective measures" to address those conditions whenever they were found. Foremost on their list of solutions was "scientific sterilization" to remove society's "worst enemy, the dregs of itself."[72]

The completed investigation confirmed many details in the picture Caldwell had painted. The elder Caldwell said that such conditions were "the result of poverty and ignorance bred through generations." His own objective in assisting the investigation was to prevent "the development of such people in the future."[73] Buttressing these lay observers' conclusions, a social worker wrote that such degraded people as those Caldwell described should be sent to institutions. Their homes should be leveled and burned, she said. Setting aside sentimentality, she declared: "These people are a cancer on society, a menace to themselves and the state; and to perpetuate their condition only increases their number."[74] Blaming the victims salved the collective conscience of Caldwell's critics. *Time* summarized the controversy and the *Augusta Chronicle* reportage, which concluded that 1 percent of east Georgia's poor were outcasts from civilization and could only be treated by sterilization. By then a bill to ensure that result had passed the Georgia House and was pending before the Georgia Senate.[75] The result of the Caldwell controversy was clear to the *Augusta Chronicle*. Although they faulted the novelist, who "laundered our dirty linen and rattled the skeletons in our closet before hundreds of thousands of people," they conceded that the episode had a constructive conclusion, because "people are aroused" to find a remedy to the social ills that had been put on display. The adult members of the 10–35 families, the "flotsam and jetsam in the sea of human misery" identified in the paper's investigation, should become wards of the state, "taken care of until they themselves pass out" of existence through death. As for their children, they should be sterilized so that "their race will be extinguished with the next generation." The newspaper noted that after many years of advocacy, working "with forward looking, patriotic Georgians," the sterilization bill had been adopted by both legislative houses and was awaiting the governor's signature, representing "the first step forward in our great social problem."[76] The bill created a state board of eugenics and allowed superintendents of state asylums to recommend candidates for sterilization. A final amendment also allowed chain gang wardens to name cases for surgery.[77]

Not everyone agreed with the legislation. The *New Republic* followed the Caldwell flap and congratulated the *Augusta Chronicle* on its "entire truthfulness" in investigating local conditions. The review proved that "there are Jeeter Lesters in the world and that something should be done to remedy the conditions under which they are forced to live."[78] Addressing rural conditions rather than the reproductive potential of the poor seemed the best course to other outside observers. But back in Georgia the *Atlanta Constitution* saluted the sterilization measure, saying it appealed to "the common sense and reason of the people." The newspaper also quoted the newly deceased Oliver Wendell Holmes Jr., whose famous epigram a decade earlier signaled Supreme Court endorsement of eugenic sterilization: "Three generations of imbeciles are enough."[79] Deep Southerners seldom accorded such tribute to the opinions of former officers of the Grand Army of the Republic.

Although the legislature had adopted the sterilization bill, in a pique of partisan revenge, Governor Eugene Talmadge refused to sign it. Turning to Adjutant General Lindley W. Camp, Talmadge said: "Lindley, you and I might go crazy some day and we don't want them working on us."[80] His quip underscored the personal rather than policy concerns prompting his veto. The sterilization bill was only one of the more than 165 other bills Talmadge vetoed that session.

Talmadge's veto "turned back the hands of the clock," declared the *Augusta Chronicle*. The veto was one of the governor's "most egregious mistakes," and it "struck a blow in behalf of illiteracy, degeneracy, imbecility, insanity and crime."[81] Surely the governor's spite was out of step with the best southern traditions and most forward-looking boosters' hopes. But in neighboring Alabama, Governor Bibb Graves also vetoed a sterilization bill that year, while to Georgia's northeast, a bill introduced by freshman South Carolina state legislator Strom Thurmond became law.[82] Sterilization, it seemed, was an issue intensely sensitive to local context.

After the veto of the Georgia sterilization law, Caldwell further inflamed local opinion when his 1935 dispatches from Georgia were collected in a book entitled *Some American People*. In the book, he condemned the plan to sterilize sharecroppers and the failure of such eugenic policies, which do not "remedy the cause of the conditions" that generate social problems. While conceding that sterilization might be appropriately applied in "certain individual cases," Caldwell rejected the measure for the "thousands of Southern tenant farmers . . . in an economic condition that demands much more than superficial thought."[83] In his later work Caldwell would repeatedly emphasize the plight of poor southern share-croppers as a situation that could be improved without resort to surgical intervention.

With a new governor in place in 1937, legislators from Caldwell's home county reintroduced the sterilization bill, which passed the Georgia House of Representatives easily.[84] Then, "without comment or debate," the Senate unanimously adopted the measure, designed "for immunization from procreation of all insane persons and habitual criminals."[85] When the legislative session ended, Governor E. D. Rivers signed the bill without fanfare, making Georgia the final state to adopt a eugenic sterilization law.[86] The *Journal of the Georgia Medical Society* applauded the law as an appropriate step to address the "constant increase in the tax burden."[87]

In 1936, photographer Margaret Bourke-White toured the South with Caldwell collecting images for another book, *You Have Seen Their Faces*.[88] The book's illustrations included actual pictures of the Bungler clan, revealed first in the journal *Eugenics*, then immortalized as fiction in *Tobacco Road*. *You Have Seen Their Faces* was condemned in Augusta as "balderdash" and "propaganda."[89]

When the stage version of *Tobacco Road* was scheduled to play in Augusta, the sheriff from Caldwell's home town of Wrens wrote to protest.[90] The play was received "with good nature and applause" in Augusta, but in Atlanta, a new municipal censorship board banned the play as "sacrilegious" and "generally vulgar and profane."[91]

In 1940, as *Tobacco Road* was ending its record-setting seven-year run on Broadway, Ira Caldwell traveled to Washington to meet government officials in an attempt to lure money for a "rehabilitation community" that would lift the inhabitants of the real "Tobacco Road" out of poverty. Revisiting the family that he first described as the "Bunglers," the elder Caldwell tried once again to better their lot. If they were provided with medicine to "wipe out their malaria" and had access to "clean and sturdy houses surrounded by a few fertile acres," they could "lift themselves out of their dilemma," Caldwell argued.[92]

By then *Tobacco Road* had found a new incarnation as a movie, and the senior Caldwell acted as tour guide to Hollywood filmmakers who visited the region looking for authentic settings for the film.[93] With an eye to a possible windfall for the local economy, the people of eastern Georgia seemed to forget their irritation with Erskine Caldwell. The Augusta Merchants Association endorsed plans to bring the world premiere of the movie to town.[94] The local paper felt the "violent controversy" had passed and there would be no protest.[95]

But the flap was hardly over. Some theater patrons were disappointed that the film had been stripped of the salty language that provided such scandal in the book and the play.[96] On the other hand, the mayor of Augusta complained that it was "the most degrading and unwholesome film" he had ever seen. Perhaps more important, it portrayed his city as a small "backwoods community" rather than the bustling metropolis of more than 100,000 souls that it now was. Although he instructed the city attorney to sue the movie producers for $500,000 for their slanderous portrayal of Augusta, Caldwell apparently did not take the threat seriously, and nothing came of the lawsuit.[97] Erskine Caldwell, known for so long as the narrator for the "unwashed, unshaved, unclothed, uneducated, and unintelligent white trash of a poverty stricken district," left the "verminous Georgia loafers" of *Tobacco Road* behind.[98]

No evidence remains of a federal response to Ira Caldwell's pleas; he died a few years later just as World War II was coming to an end, with the country in the middle of its long climb out of the Great Depression.[99] During the next three decades, some 3,300 Georgians endured surgery under Georgia's sterilization law. Though it was in place only 33 years, in a tally of surgical statistics, Georgia ranks fifth among all states in the total number of eugenic operations. Between 1950 and 1960, approximately 200 involuntary operations occurred every year at the state's mental hospital. In 1959 a Pulitzer Prize–winning investigative series revealed that sterilizations at Georgia's Central State Hospital had been performed routinely on any patient of childbearing age and that the surgeries were often done not by a doctor but by a nurse. An advisory committee later recommended that all eugenic sterilizations be discontinued.[100]

In 1966, new legislation was written sanctioning voluntary sterilization for purposes of birth control for married couples.[101] Additional
amendments to those provisions eventually led to elimination of the law
for eugenic sterilization, which was repealed in 1970.[102]

In 2007, the Georgia Senate passed a resolution of "profound regret"
for eugenic sterilization legislation. The resolution leveled blame for passage of the 1937 measure at "Darwinian principles" along with "progressive" academicians, scientists, politicians, and newspaper editors; those
with "religious" objections were credited with opposing it.[103] No mention
was made of Erskine Caldwell or the Bunglers of *Tobacco Road*.

NOTES

1. Martin Pernick, *The Black Stork: Eugenics and the Death of "Defective" Babies in
American Medicine and Motion Pictures since 1915* (New York: Oxford University Press,
1996), 54.

2. Some studies that have highlighted eugenic themes in literature are Donald
J. Childs, *Modernism and Eugenics: Woolf, Eliot, Yeats, and the Culture of Degeneration*
(Cambridge: Cambridge University Press, 2001); Daylanne K. English, *Unnatural Selec-
tions: Eugenics in American Modernism and the Harlem Renaissance* (Chapel Hill: Uni-
versity of North Carolina Press, 2004); Sylvia J. Cook, "Modernism from the Bottom
Up," in *Reading Erskine Caldwell: New Essays*, ed. Robert L. McDonald (Jefferson, N.C.:
McFarland, 2006); and Karen Ann Keely, "The Pure Products of America: Eugenics and
Narrative in the Age of Sterilization" (Ph.D. diss., UCLA, 1999).

3. Among the most cited popular fiction that includes references to eugenics
are Sinclair Lewis, *Main Street* (1920), *Babbitt* (1922), and *Arrowsmith* (1925); F. Scott
Fitzgerald, *The Great Gatsby* (1925); and Ernest Hemingway, *Torrents of Spring* (1926).
Fitzgerald began his references to eugenics as a student at Princeton, where he wrote
the lyrics for a ditty entitled "Love or Eugenics" for a Triangle Club production. Banned
from participating in the play due to poor academic performance, Fitzgerald appeared
in drag, dressed as a chorus girl. See F. Scott Fitzgerald, *Fie! Fie! Fi-Fi! A Facsimile of
the 1914 Acting Script and Musical Score* (Columbia: University of South Carolina Press,
1996).

4. This connection is suggested, for example, in Dan B. Miller's biography *Erskine
Caldwell: The Journey from Tobacco Road* (New York: Alfred A. Knopf, 1995), 219.

5. Edward J. Larson, *Sex, Race, and Science: Eugenics in the Deep South* (Baltimore:
Johns Hopkins University Press, 1995).

6. Karen A. Keely, "Poverty, Sterilization, and Eugenics in Erskine Caldwell's
Tobacco Road," *Journal of American Studies* 36 (2002): 23–42; Betsy L. Nies, "Defending
Jeeter," in *Popular Eugenics: National Efficiency and American Mass Culture in the 1930s*,
ed. Susan Currell and Christina Cogdell (Athens: Ohio University Press, 2006), 120–39.

7. W. E. Gonzales, "National Corn Exposition and Its Big Meaning," *Atlanta Constitution*, January 25, 1913; "Need of Child Welfare Exhibit Is Arousing Much Interest,"
Atlanta Constitution, September 21, 1913.

8. "Eugenics for Atlanta Is Strong Prediction," *Atlanta Constitution,* November 19, 1914.

9. A. T. Spaulding Jr., "How Mendel's Laws Discredit Eugenics," *Atlanta Constitution,* March 9, 1914. The Better Babies gave way to the more formally organized Fitter Family contests run by the American Eugenics Society. In 1924, an important year for Georgia eugenics, the family of a school principal took the gold medal at the Fitter Families contest at the Georgia State Fair. See "School Principal and Family Take Fair Top Honors," *Savannah Press,* November 6, 1924.

10. "Britain Is Opposed to Eugenics in Schools," *Atlanta Constitution,* January 9, 1914.

11. "No Eugenics in Schools," editorial, *Atlanta Constitution,* January 11, 1914.

12. Paul Jones, "Governor and Mayor Are Lauded by Sunday," *Atlanta Constitution,* November 5, 1917.

13. "Damaged Goods," *Atlanta Constitution,* January 8, 1916.

14. "Walter Andrews Now Believer in Eugenics," *Atlanta Constitution,* January 2, 1916.

15. "Eugenics Bill Bitterly Fought by Senator Kea," *Atlanta Constitution,* July 13, 1920.

16. "Kill Marriage Bill after Two Day Fight: Senators Defeat Bill Requiring Certificate of Health before Issuance of License," *Atlanta Constitution,* July 14, 1920.

17. "Eugenics Bill Rejected," *Washington Post,* August 2, 1921; "Kennedy Offers Eugenics Bill," *Atlanta Constitution,* July 7, 1923; "Legislators Open Determined Plan to Clear Slate," *Atlanta Constitution,* July 20, 1924.

18. *Journal of the Medical Association of Georgia* 17 (1928): 308–309.

19. Ellis Parker Butler, "Suzanne on Race Tinkering," *Atlanta Constitution,* June 7, 1914.

20. "Racial Integrity as a National Question," editorial, *Atlanta Constitution,* February 7, 1907.

21. "Racial Marriage a National Problem," editorial, *Atlanta Constitution,* August 21, 1908.

22. "The Two Go Hand in Hand," editorial, *Atlanta Constitution,* June 21, 1914.

23. "The Public Health Bill," *Atlanta Constitution,* July 12, 1914.

24. Paul A. Lombardo, "Miscegenation, Eugenics & Racism: Historical Footnotes to *Loving v. Virginia,*" *University of California, Davis Law Review* 21 (1988): 422–52.

25. Walter A. Plecker to Rev. Wendell White, May 10, 1924, E. S. Cox Papers, Duke University [hereafter Cox Papers].

26. James C. Davis to John Powell, May 25, 1925, John Powell Papers, University of Virginia [hereafter Powell Papers].

27. "Appeal Is Made to Keep Blood of Whites Pure," *Atlanta Constitution,* February 13, 1924.

28. Powell to Davis, May 30, 1925, Powell Papers.

29. Sam W. Small, "Wide Interest Is Aroused in Racial Integrity Bill," *Atlanta Constitution,* June 28, 1925.

30. Davis to Powell, August 22, 1927, Powell Papers.

31. Dr. G. I. Garrard, cited in Peter G. Cranford, *But for the Grace of God: The Inside Story of the World's Largest Insane Asylum, Milledgeville!* (Augusta, Ga.: Great Pyramid Press, 1981), 68.

32. "Doctors of Georgia Plan Bill for Sterilization of the Criminally Insane," *Atlanta Constitution*, April 18, 1913; "A Worthy Measure," *Atlanta Constitution*, May 30, 1913. One out-of-state advocate of eugenic surgery described the operation as the solution for the "vexing negro problem." Since more than 50 percent of all convicted criminals are blacks, he said, sterilization of the prison population would necessarily apply primarily to that group. "Declares Sterilization Is Solution of Negro Problem," *Atlanta Constitution*, April 20, 1913.

33. W. L. Champion, "Sterilization of Confirmed Criminals, Idiots, Rapists, Feeble-Minded, and Other Defectives," *Journal of the Medical Association of Georgia* 3 (1913): 112–14, at 113.

34. "Remaking the Universe," *Atlanta Constitution*, June 28, 1914.

35. W. B. Hardman, "The Medical Gospel of the Twentieth Century," *Journal of the Medical Association of Georgia* 4 (1914): 71–75.

36. "Anderson Urges Mentality Tests," *Atlanta Constitution*, March 30, 1919.

37. "The Feebleminded," *Atlanta Constitution*, June 2, 1919.

38. "The Cost of Feeblemindedness," *Atlanta Constitution*, May 29, 1922.

39. "Financial Aspect of Feeblemindedness," *Atlanta Constitution*, May 31, 1922.

40. "What Feeblemindedness Means to the Community and What the Community Can Do to Protect Itself," *Atlanta Constitution*, June 4, 1922.

41. "State Sanitarium Short of Funds," *Augusta Chronicle*, April 18, 1929.

42. "State Legislation," *Journal of the Medical Association of Georgia* 21 (1932): 335.

43. "Sterilization of Epileptics Urged," *Augusta Chronicle*, May 14, 1929.

44. I. S. Caldwell, "The Eugenics Question," *Augusta Chronicle*, May 20, 1929; I. S. Caldwell, "Are We Going Up or Down?" *Augusta Chronicle*, December 5, 1929.

45. Miller, *The Journey from Tobacco Road*, 125.

46. I. S. Caldwell, "The Bunglers: A Narrative Study in Five Parts," *Eugenics* 3 (1930): 202–10 (part 1); 247–51 (part 2); 293–99 (part 3); 332–36 (part 4); and 377–83 (part 5). Comment on "quicksands of ignorance" on 295.

47. Ibid., 207.

48. Ibid., 332.

49. Ibid., 383.

50. R. J. Terry, "Robert Bennett Bean, 1874–1944," *American Anthropologist*, n.s., 48 (1946): 70–74; Miller, *The Journey from Tobacco Road*, 77. (Miller mistakenly thought Caldwell had studied with Bean's son, William.)

51. Erskine Caldwell, *Call It Experience: The Years of Learning How to Write* (New York: Duell, Sloan and Pierce, 1951), 34–35.

52. Erskine Caldwell, "Georgia Cracker," *Haldeman-Julius Monthly* 4 (November 1926): 39–43, at 41.

53. Caldwell, *Call It Experience*, 81.

54. See, for example, "Erskine Caldwell: Phenomenon," *Gastonia (N.C.) Daily Gazette*, March 2, 1935 ("a notch above the filthy little sex booklets"; "nice picking for the literary rag man"; designed for "the entertainment of sex-greedy filth scavengers").

55. "Atlanta Doctors to Drive for Sterilization," *Atlanta Constitution*, February 4, 1934.

56. "Report of the Committee on Public Policy and Legislation," *Journal of the Medical Association of Georgia* 23 (1934): 304.

57. "Legal Sterilization Advocated in Address by Emory Biologist," *Atlanta Constitution*, March 4, 1934; "Sterilization Is Urged by Biologist to Protect Unborn from Unfitness," *Atlanta Constitution*, April 14, 1934.

58. "Sterilization by State Urged by Georgians," *Atlanta Constitution*, April 29, 1934.

59. "Lindsey Urges Sterilization as Insanity Ban," *Atlanta Constitution*, March 29, 1934.

60. "Sterilization in California," *Atlanta Constitution*, April 27, 1934.

61. "Sterilization Act Expected to Come before Assembly," *Atlanta Constitution*, November 25, 1934.

62. Erskine Caldwell, "Georgia Poverty Swept, Says Caldwell," *New York Post*, February 18, 1935.

63. Erskine Caldwell, "Landowners Find Help Gives Them Edge on Laborers," *New York Post*, February 19, 1935.

64. Erskine Caldwell, "Georgia Land Barons Oust Dying Girl and Her Father," *New York Post*, February 20, 1935.

65. Erskine Caldwell, "Starving Babies Suckled by Dog in Georgia Cabin," *New York Post*, February 21, 1935.

66. "Erskine Caldwell, Noted Author, Is Visitor in Augusta," *Augusta Chronicle*, December 23, 1933.

67. "Caldwell Story Called Untrue; Probe Ordered," *Augusta Chronicle*, March 5, 1935.

68. "What Will the Good People of Jefferson County Say of This?" *Augusta Chronicle*, March 4, 1935.

69. Letter to the Editor, "Share Cropper Conditions Denied," *New York Post*, March 6, 1935.

70. "Farmers," *Time* 25 (March 4, 1935): 13.

71. Letters: "Share Croppers," "Crackpot Version," "Caldwell's Hog Waller," and "Father," all in *Time* 25 (March 25, 1935): 5–8.

72. "Investigation Is Made to Determine Basis Caldwell Had for All His Writings," *Augusta Chronicle*, March 10, 1935.

73. "Erskine Caldwell," *Augusta Chronicle*, March 14, 1935.

74. "From a Social Worker," *Augusta Chronicle*, March 17, 1935.

75. "Along Tobacco Road," *Time* 25 (March 25, 1935): 59–61.

76. "The Caldwell Issue and Long-Range Rehabilitation," *Augusta Chronicle*, March 24, 1935.

77. "House Votes Sterilization," *Atlanta Constitution*, March 7, 1935. On chain gangs, see also *Augusta Chronicle*, March 8, 1935.

78. *New Republic*, March 27, 1935, 173.

79. "Sterilization Bill Passes the House," *Atlanta Constitution*, March 9, 1935.

80. "Talmadge Vetoes Bill to Sterilize Criminals, Insane," *Atlanta Constitution*, March 27, 1935.

81. "The Governor's Veto of the Sterilization Bill," *Augusta Chronicle*, March 27, 1935.

82. "Sterilization Bill Is Vetoed by Graves," *Atlanta Constitution*, June 26, 1935; "Sterilization Okayed for Carolina Insane," *Atlanta Constitution*, May 19, 1935.

83. Erskine Caldwell, *Some American People* (New York: Robert M. McBride, 1935), 236.

84. "Vote Sterilization of State's Insane," *Savannah Press*, February 9, 1937.

85. "Sterilization Bill Passed," *Atlanta Georgian*, February 19, 1937.

86. "Sterilization Bill Accorded Governor's OK," *Atlanta Journal*, February 24, 1937.

87. Avary M. Dimmock, "Human Sterilization," *Journal of the Medical Society of Georgia* 26 (August 1937): 423–25.

88. Erskine Caldwell and Margaret Bourke-White, *You Have Seen Their Faces* (New York: Modern Age Books, 1937).

89. "Half-Truths Worse Than Lies," *Augusta Chronicle*, March 28, 1938.

90. "Opposes Showing of 'Tobacco Road,'" *Augusta Chronicle*, October 16, 1938.

91. "Augustans Accept 'Tobacco Road' with Good Nature and Applause," *Augusta Chronicle*, November 18, 1938; "New Board Bans 'Tobacco Road,'" *Augusta Chronicle*, November 22, 1938.

92. William Gober, "Senior Caldwell Hopes to Rehabilitate Tobacco Road," *Augusta Chronicle*, August 11, 1940.

93. Joel Huff, "Hollywood Trio Here Inspecting 'Tobacco Road,'" *Augusta Chronicle*, November 5, 1940.

94. "City Briefs," *Augusta Chronicle*, January 1, 1941.

95. "No Longer an Issue," *Augusta Chronicle*, March 2, 1941.

96. Harry Gage, "Augustans Are Disappointed in 'Tobacco Road' Picture," *Augusta Chronicle*, March 4, 1941.

97. "Augusta Mayor Lambasts Makers of 'Tobacco Road,'" *Augusta Chronicle*, March 6, 1941; "Work to Begin on Suit against 'Tobacco Road,'" *Augusta Chronicle*, March 15, 1941.

98. George Tucker, "New York," *Augusta Chronicle*, June 19, 1941.

99. "Rev. I. S. Caldwell Dies in Columbia," *Augusta Chronicle*, August 18, 1944.

100. A series of letters from the former legal counsel to the Medical Association of Georgia clarify the events that led to passage of the voluntary sterilization law as well as the eventual repeal of the Georgia eugenic sterilization act. See John. L. Moore Jr. to Julius Paul, October 10, 1968, November 4, 1968, and February 22, 1972, copies in possession of the author. See also Gayle White, "Involuntary Sterilization in Georgia: Why Did It Happen?" *Atlanta Journal-Constitution*, February 4, 2007.

101. "Voluntary Sterilization Act," *Georgia Acts* 1966, p. 453.

102. "Voluntary Sterilization Act," *Georgia Acts* 1970, p. 683. The 1970 law retained provisions for sterilization of people who are "irreversibly and incurably mentally incompetent" and could not raise children safely. That law was successfully challenged and declared unconstitutional for failing to provide procedural and evidentiary protections for those subject to sterilization. See *Motes v. Hall County Department of Children and Family Services*, 251 Ga. 373, 1983. The faulty provisions were amended, and current Georgia law continues to allow sterilization of the mentally disabled following a court order.

103. *Georgia Senate Resolution 247* (2007), "Expressing profound regret for Georgia's participation in the eugenics movement in the United States." For details of the resolution and the political machinations that accompanied it, see Paul A. Lombardo, "Disability, Eugenics, and the Culture Wars," *St. Louis University Journal of Health Law and Policy* 2 (2009): 57–79.

FOUR

"Quality, Not Mere Quantity, Counts": Black Eugenics and the NAACP Baby Contests

GREGORY MICHAEL DORR AND ANGELA LOGAN

In the past decade, Americans have rediscovered their nation's eugenic past. Governors and legislators in Virginia, Oregon, North Carolina, South Carolina, California, Indiana, and Georgia have made proclamations of apology, adopted resolutions of regret, and voiced general repudiations of past eugenic laws and other efforts to purify the citizenry by skimming allegedly antisocial traits from the gene pool through practices such as compulsory sterilization. Popular accounts of eugenic history in the media have depicted racist white elites abusing "unfit" populations—specifically the mentally retarded, lower-class women, and racial and ethnic minorities. At least two films on the U.S. eugenics movement have been made from this perspective.[1]

These media depictions were consistent with the bulk of scholarship on the eugenics movement. For approximately 30 years, most historians of eugenics have focused on prominent white scientists, psychologists, and eugenic propagandists. More recently, attention has turned to "second tier" white eugenicists and state-level studies of the pro-eugenics rank and file, illustrating the pervasive nature of eugenic ideology across white American culture. The influence of hereditarian ideas within the African American community has received much less attention. Most studies position African Americans as the targets of eugenic control and repression, or as vocal—if disempowered and ignored—critics of eugenics. These accounts strip black historical actors of their agency and oversimplify the American eugenics movement.[2]

Hereditarianism and eugenics held strong appeal for various segments of the African American community. While very few African Americans accepted the claims made by black polemicists like William Hannibal Thomas—who posited the hereditary inferiority of his own race—many blacks endorsed the Nation of Islam's (NOI) proto-genetic-engineering eschatology. Followers of NOI leader Elijah Muhammad believed the evil breeding experiments of Yakub, the "big head scientist," resulted in the white race and its political domination of blacks. Other African Americans flocked to Marcus Garvey's Universal Negro Improvement Association (UNIA), adopting his zeal for maintaining black racial purity and his advocacy of race separation that echoed the eugenically tinged "racial integrity" programs propounded by segregationist whites. Yet all three of these thinkers, their organizations, and the counternarratives they built to combat white scientific racism emerged after an earlier—and to some degree subtler—form of African American hereditarianism. In fact, the theories of Thomas, Muhammad, and Garvey each responded as much to the shortcomings of this earlier class-biased and race-neutral black eugenics as they did to the racist provocations of white eugenicists.[3]

This essay maps the broad contours of what might be called "assimilationist" eugenics—a perspective that viewed racial difference as insignificant, but adopted more fundamental eugenic notions about distinctions between "fit" and "unfit" people. This theory of human breeding, applicable to all people regardless of race, could appeal to blacks. Assimilationist eugenics grew out of older traditions of African American racial egalitarianism. Popularized by public intellectuals like W.E.B. Du Bois, and inculcated among the black elite by educators like Thomas Wyatt Turner, assimilationist eugenics informed the thinking of many upper-class blacks during the first third of the twentieth century. Contrasting Du Bois and Turner's assimilationist eugenics to Thomas's arguments for black inferiority, Muhammad's quasi-scientific Afrocentrism, and Garvey's racial purity crusade underscores the intellectual thread woven through all four ideologies. All of these men believed in eugenics' central dogma that (a) human beings could be sorted into the relatively "fit" and "unfit" and (b) society as a whole could be improved by ensuring the propagation of the fit and reducing procreation among the unfit. While they differed over what attributes characterized fitness and determined membership in that valorized group, they agreed that concrete programs could translate thought into action, uplifting the race.[4]

The second part of this essay analyzes the National Association for the Advancement of Colored People's baby contests, a key component of the assimilationist eugenic agenda. More than infant beauty pageants, these competitions blended ideology and activism. Baby contests popularized race-neutral, class-inflected eugenic ideals, drafting these hereditarian principles into the NAACP's fight against the fruits of white eugenics: segregation, antimiscegenation law, and lynching. The baby contests highlight how middle-class African Americans co-opted and inverted the repressive ideas and programs of "mainline" racist white eugenicists, recasting them into arguments and policies that supported class-inflected social justice through African American racial uplift.[5]

ASSIMILATIONIST EUGENICS

ORIGINS

From the moment Sir Francis Galton coined the word *eugenics* in 1883, the notion of scientifically improving humanity through better breeding captivated modern thinkers. Whether advocating increased procreation among the "fit"—so-called positive eugenics—or demanding negative eugenic interventions like immigration and marriage restriction, sterilization, and segregation to reduce the propagation of the "unfit," theorists remained convinced that eugenics promised utopian improvement for humanity. A host of social problems—alcoholism, criminality, pauperism, prostitution, tuberculosis, venereal disease, and above all "feeblemindedness"—might be eradicated by preventing the birth of those genetically predisposed to these maladies. Definitions of fitness and unfitness corresponded to native-born, white Americans' racial, class, and ethnic prejudices, underscoring the eugenics movement's overlapping scientific and cultural imperatives. But despite affinities between eugenics and white supremacy, black folk recognized promise and power in the new science, too.

Any discussion of African Americans' eugenic beliefs must be placed against the backdrop of white and black support for "popular eugenics": the ambient, if nebulous, belief in the existence of "fit" and "unfit" babies blessed or cursed with "good" or "bad" physical and behavioral characteristics. Popular eugenics—like the more organized, white-dominated mainline movement itself—swept over American society during the 1910s and 1920s. It is hardly surprising that white *and* black Americans often analogized human heredity to the domestication and "improvement" of animals and plants—a comparison fostered by direct experience with stockbreeding and farming. Census data indicated that until 1920 the majority of the U.S. population lived in rural areas; 90 percent of African Americans lived in farm settings until the Great Migration took over a million blacks from the rural south to industrial jobs, predominantly in northern and eastern cities. Familiarity with the basic concepts of husbandry and horticulture demystified eugenic concepts; attendance at (and enjoyment of) county and state fairs' stock, crop, and baking competitions made "better babies" and eugenic "Fitter Families" contests seem more like an entertaining diversion than a program for social control.[6]

Like their white counterparts, many African Americans had heard the conventional wisdom, or concluded from direct experience, that "fit" and "unfit" individuals existed. Just looking around, they saw evidence of imperfection in their neighbors who were mentally slow, physically disfigured, or morally lax. Many Americans, regardless of race, became interested in finding a "fit" mate and having "fit" children—both to enhance their progeny's life prospects and to fulfill their social responsibility to improve the population. By emphasizing familiar premises of hereditary determinism and genetic improvement—rather than black inferiority (Thomas), deranged black scientists and genetically engineered white devils (NOI), or light-skinned "race traitors" and the need for black "racial purity" (UNIA)—assimilationist eugenic theorists wedded popular ideas to more esoteric scientific concepts, appropriating the power of science to serve integrationist political ends. In constructing a "scientific" counternarrative to combat the "scientific" racism of mainline white eugenicists, assimilationist eugenicists sought to "fight fire with fire," turning racist theory on its head.[7]

THEORY

W.E.B. Du Bois led the intellectual assault on the hereditarian doctrines undergirding the mainline eugenics movement. Du Bois's efforts to define race as a historical entity, his attempts to debunk purported racial differences in intelligence, and his steadfast objection to racial supremacy have often been cited as evidence of his hostility toward hereditary determinism and eugenics. Although Du Bois railed against white supremacists who used the "objectivity" of "eugenic science" to mask their racism, a close reading of Du Bois's work reveals his nuanced deployment of hereditarian ideas in his social theorizing. For Du Bois, the complexity of human development exceeded the "either-or" simplisms of heredity versus environment, nature versus nurture. He maintained a healthy respect for the notion of genetic fitness, even as he remained mindful of environment's power in shaping human destiny. Du Bois's intellectual formulation of the "Talented Tenth"—those African Americans he believed destined to lead the larger black population to equality—was shot through with this sophisticated hereditarianism.[8]

Rather than debating the relative influence of "nature" or "nurture," Du Bois's hereditarianism described an arc from racial essentialism to intellectual elitism. In his controversial 1897 address "The Conservation of Races," he acknowledged that "there are differences—subtle, delicate, and elusive, though they may be—which have silently but definitely separated men into groups." Du Bois believed that most of these observable, socially salient "racial" differences resulted from the influence of historical and cultural context (environment). Blacks' destiny, according to Du Bois, was neither biological extinction (through race war) nor "absorption by the white Americans" (through either acculturation or interbreeding). Instead, "it is our duty to conserve our physical powers, our intellectual endowments, our spiritual ideals" through "race organization, by race solidarity, by race unity," in other words, through efforts to improve the sociopolitical environment of African Americans. Yet, whatever the power of environment, African Americans' experience was also shaped by heredity.[9]

Political and cultural solidarity needed to be augmented by eugenic biology: proper breeding. Only procreation among fit blacks would eradicate the race's "heritage of moral iniquity," an infirmity that Du Bois described as the "hereditary mass of corruption from white adulterers" that was genetically instilled in some mixed-race children during slavery and after and then, thanks to Jim Crow, concentrated in the black population through interbreeding. Only a dual assault, through rearing and breeding, could meliorate slavery's social and biological legacy. Unlike white eugenicists who rhetorically acknowledged the coequal importance of heredity and environment before privileging heredity's effect on individuals and groups, Du Bois affirmed the nature/nurture dialectic and tracked its variable effect on African American destiny.[10]

In his famous 1903 essay, "The Talented Tenth," Du Bois wrote, "the Negro race, like all races, is going to be saved by its exceptional men." This "revolutionary group of distinguished Negroes . . . persons of marked ability" were "leaders of a Talented Tenth, standing conspicuously among the best of their time" who represented social, political, and biological salvation. Using language resonant with medical and eugenic ideas of purity and superiority and emphasizing it with capital letters, Du Bois asserted that African Americans must cultivate "the Best of [their] race that they may guide the Mass away from the contamination and death of the Worst in their own and other races." Underscoring the identity of class and quality, Du Bois acknowledged that environmental considerations conditioned the ability of the fit to improve the inferior: "From the very first, it has been the educated and intelligent of the Negro people that have led and elevated the mass, and the sole obstacles that nullified and retarded their efforts were slavery and race prejudice." Slavery and racism, however, were more than simply environmental constraints. Du Bois averred that "slavery was but the legalized survival of the unfit and the nullification of the work of natural internal leadership." Despite slavery's dysgenic effects, the Talented Tenth had managed to survive. Du Bois charged these elite African Americans with rising and pulling "all that were worth the saving up to their vantage ground." Not every black should be saved—only those "fit" ones whose salvation would improve and advance the race by expanding the ranks of the fit within the black population.[11]

Du Bois's incipient (and ultimately short-lived) political and biological isolationism evolved into his belief in the essential biological equality of the races. As early as 1899, Du Bois had identified a "submerged tenth" among blacks comprised of the "lowest class of criminals, prostitutes, and loafers"—exactly the sort of people white eugenicists would excoriate as the "shiftless, ignorant and worthless class of antisocial whites" some 25 years later. By 1904, Du Bois argued that "the Negro races are from every physical standpoint full and normally developed men [that] show absolutely no variation from the European type sufficient to base any theory of essentially human difference upon." While the races were biologically equal, that did not mean that there were not fit and unfit individuals within each race—merely that the races, as groups, defied hierarchical organization. Du Bois wrote that "the Negroes have their degenerate types in the dwarfs and Hottentots—so have the Europeans; they have their mixed types of all degrees and kinds of mixture—so have the Europeans. But it is an unproved and to all appearance an unprovable thesis that the physical development of men shows any color line below which is black pelt and above the white." That *individuals* might be ranked according to innate ability, however, was both possible and, from Du Bois's self-proclaimed intellectual vantage point, self-evident.[12]

In 1910 Du Bois declared that "there are human stocks with whom it is physically unwise to intermarry, but to think that these stocks are all colored or that there are no such white stocks is unscientific and false." Perhaps thinking of his own multiracial heritage, Du Bois argued that racial intermarriage was "not necessarily undesirable and race blending may lead, and often has led, to new, gifted, and desirable stocks and individuals." Individual fitness, not race, was the yardstick for measuring biological worth. This premise anchored Du Bois's conception of the Talented Tenth as "a carefully bred, selected, and trained elite."[13]

Just as racial harmony could result, in the near term, from the leadership of and cooperation among the "best" whites and blacks, long-term African American advance depended upon the prolific reproduction of the Talented Tenth. Social improvement and political equality would be ensured as the multiplying fit gradually replaced the unfit. Instead of regressing toward the mean, the future African American population would be socially and biologically uplifted to the status of the present Talented Tenth of *all* races. Assimilationist eugenics, like mainline white eugenics, represented a call for biological meliorism to serve sociopolitical ends.[14]

The hereditary superiority of the Talented Tenth appears throughout Du Bois's writing, but he never neglects environment. He argued that all environmental legacies of slavery (like disfranchisement, Jim Crow, racism, and what today might be labeled "environmental racism"—the shunting of blacks to unsanitary urban ghettos and parasite-infested rural margins) must be removed to ensure the maximum attainment of black potential. But heredity also deserved its due. Not all inequality stemmed from the environment. In *Souls of Black Folk,* Du Bois confirmed his belief in "the rule of inequality:—that of the million black youth, some were fitted to know and some to dig." He made this inborn individual fitness the basis of his call for education according to inborn ability, regardless of race, class, gender, or other cultural considerations. He pronounced in 1912 that, to achieve their rightful leadership role, "the Talented Tenth must be created based upon the commitment of honest colored men and women not to bring aimless rafts of children into the world, but, as many as, with reasonable sacrifice, we can train to largest manhood." Acknowledging the coequality of heredity and environment, Du Bois further declared that African Americans, "having procreated reasonably, must then seek out colored children of ability and ... find that Talented Tenth (in the next and subsequent generations) and encourage it by the best and most exhaustive training in order to supply the Negro race and the world with leaders, thinkers and artists." Reasonable, eugenic breeding and euthenic, environmental conditioning would save the race.[15]

Environmental influences (training, social background) had to be matched with hereditary ability (genius, inborn aptitudes). In the August 1922 issue of the *Crisis*, Du Bois further articulated his eugenic goals: "The Negro has not been breeding for an object; therefore, he must begin to train and breed for brains, for efficiency, for beauty." He enlarged on this theme in October, when he wrote, "Birth control is science and sense applied to the bringing of children into the world, and of all who need it we Negroes are the first." Du Bois, like many of his white eugenicist contemporaries, felt that "we in America are becoming sharply divided into the mass who have endless children and the class who ... have few or none." Surely upper-class folk could provide their children with enriched environments and experiences. Their ability to do so, however, was predicated on their success, which was itself grounded on their inborn fitness. Fitness was both a cause and signifier of their upper-class status: a positive hereditary legacy they would pass on to their children.[16]

Du Bois sought the widest audience for these ideas. The NAACP's house organ, the *Crisis*, under Du Bois's founding editorial leadership, became the mouthpiece for assimilationist eugenics and a "eugenic 'family album,' a visual and literary blueprint for the ideal, modern black individual." The *Crisis* quoted psychologist Albert Sidney Beckham's opinion: "Eugenics will improve the Negro of the future."[17] The magazine also advanced assimilationist eugenics in subtle ways. The "Men of the Month" issues, for instance, were devoted to "establishing the 'college-bred,' middle-class, urban, intellectual man as the authentic representative of an ideal racial family." This "New Negro" man was intended to serve as "visual confirmation of the genetic uplift of the race." The magazine, however, sought more than tallying examples of black "manhood" to counter white claims of innate African American "savagery."[18]

The October "Children's Number," which began the year of the publication's inception and ended the year Du Bois resigned as editor, featured stories, editorials, and pictures of youth and babies. The issue emphasized the importance of uplifting the race through increasing and promoting the Talented Tenth's progeny. Compilations of outstanding achievements by children and young adults served two purposes: they countered "racist representations of blacks by the white press" and kept black folk updated on and encouraged by recent achievements.

For its first eight years, the Children's Number promoted the achievements of select children, chosen to represent the proper confluence of environment and heredity. Du Bois conceded, "True, these are selected children," but they were harbingers of "a large and larger class of well-nourished, healthy beautiful children among the colored people," the eugenic progeny of the fit. The 1927 edition listed "five exceptional Negro children" not by name but by intelligence quotient—implying the children's hereditary fitness. Although Du Bois decried white eugenicists' use of intelligence testing to rank races, he acknowledged heredity's role in shaping intellect and used one of the eugenicists' most prized tools—IQ tests—to counter scientific racists' accusations of black intellectual inferiority.[19] He and his supporters invoked heredity in support of the eugenic breeding of the Talented Tenth among *all* races.

In 1932, Du Bois contributed an essay on birth control to Margaret Sanger's *Birth Control Review*. Du Bois accepted the conventional eugenic wisdom that "the more intelligent class" of all races used contraception. But because the unfit cohort was disproportionately large among blacks, "the increase among Negroes, even more than the increase among whites, is from that part of the population least intelligent and fit, and least able to rear their children properly." He complained that African Americans "must learn that among human races and groups, as among vegetables, quality and not mere quantity really counts." If his advice was followed, "In time efficiency and brains are going to be well-bred in the American Negro race." Du Bois adhered so strongly to this notion that he allowed this article to be reprinted, unchanged, in 1938. In principle, Du Bois assumed that distinctions in ability could be scientifically and objectively measured and eugenically reproduced not between the races, implying a permanent barrier, but among the races. White-dominated culture and science, however, blinded by prejudice, ignored black merit and biological potential and sought to limit fitness to the white race. Other black intellectuals shared Du Bois's meritocratic, antiracist view and worked hard to popularize it.[20]

TEACHING

Dr. Thomas Wyatt Turner adhered to a more explicitly hereditarian worldview than Du Bois. The son of former slaves, a devout Catholic, a charter member of the NAACP, and a founder of its Washington, D.C., chapter, Turner maintained three identities often believed incompatible with support for eugenics. Nevertheless, after graduating from Howard University in 1901, Turner taught biology and eventually eugenics at Tuskegee, Howard, and Hampton universities. He first encountered eugenics during the summer of 1904, when he studied at the Long Island Biological Laboratory under Dr. Charles Benedict Davenport, the dean of American eugenics and a determined, if genteel, racist. Turner's career would exemplify how African American intellectuals, including scientists, could exploit eugenics' ideological flexibility to harmonize their racial, religious, and reformist beliefs.[21]

Turner taught generations of students about the power of heredity in human affairs. After a decade (1901–12) at Tuskegee, Turner joined the Howard faculty and began teaching eugenics in 1913. Turner used Paul Popenoe and Roswell Johnson's mainline classic, *Applied Eugenics,* as his text. He also assigned Davenport's famous essay, "Eugenics and Euthenics," suggesting Turner's effort to equilibrate the influence of heredity and environment. Still, examining his Sex Hygiene students in 1915, Turner asked them to "Define Eugenics. Explain how society may be helped by applying eugenic laws." A 1920 exam asked almost the same question. Turner expected his students to learn and adopt the major hereditarian tenets of eugenic improvement.[22]

Like Du Bois and other black intellectuals, Turner identified the Talented Tenth—people like himself and his students—as the eugenically fit who should procreate to improve the black population. Turner agreed with Du Bois that the "best" blacks were every bit as "fit" as the "best" whites, ignoring race and accepting the class biases of white eugenicists. Asserting that educational and economic achievement signified the power of heredity allowed African American leaders to consider themselves "racial exemplars" in the most essential biological terms. Their intellectual and socioeconomic success became a self-justifying prescription and prognosis for black eugenic salvation.

Assimilationist eugenic theory thus found a ready partner in the period's most widespread, class-based, intraracial reform ideology, known colloquially as "racial uplift." This impulse was characterized by the noblesse oblige of the Talented Tenth and encapsulated in the National Association of Colored Women's motto: "Lifting as We Climb." Turner argued that if African Americans used biology and eugenics to "uplift" the race to meet white norms on the fundamental biological level, then cultural disparities would vanish.

Eugenical uplift would undermine white appeals to biology as justification for black inequality, taking heredity out of the interracial nature/nurture debate. Forced to admit human beings' fundamental genetic similarities, whites would have to concede that inequity resulted from the environmental constraints of racism, not genetic essence. Assimilationist eugenic uplift, in this context, provided a method for regaining black citizenship.[23]

Turner taught mainline beliefs regarding the "inferiority" of the feebleminded, even as he reconfigured these notions to create a scientifically inflected liberation ideology for African Americans. He declared the feebleminded "incapable of competing on equal terms" with normal people. Most white eugenicists contended that this inability to compete also characterized black/white *group* relationships, implying that all blacks were inherently feebleminded. Turner rejected this suggestion, asserting that "defectiveness is not a reversion but direct inheritance," debunking the white shibboleth that blacks were devolving to "savagery," and distancing "normal" fit blacks from "unfit," "retarded" blacks and whites. Defectives, according to Turner, existed in and should be eliminated from both races—for the good of the entire human race. Eliminating the unfit would foster the interracial collaboration among elites that Du Bois foretold in *Souls*.[24]

For Turner and mainline eugenicists, all sound public policy must stem from a biological understanding of life. Social reform must "look to the improvement of the individual as well as to the improvement of the race," it must "make society better by working upon the individual units of society," and it would ideally "aim not only at ameliorating the conditions of life but also at bettering life itself." Just as white eugenicists distrusted sociology's environmental emphasis, Turner believed that the sociologist "must be primarily a biologist," if he hoped to achieve substantive reforms.[25]

Ethical public policy and "sound citizenship" depended upon merging the scientific method with civics instruction. Turner averred that contemporary Americans were "nearer the goal of universal brotherhood" than the founders, "because the pursuit of science has developed a larger sympathy among men, by teaching them that they are truly of one flesh, with a common parentage." For Turner, biology explained humanity's phenotypic distinctions in terms of the underlying organic similarity of all human beings.[26]

Turner recast racial disparities into an associational rather than a segregationist framework. Inverting white eugenicists' "race suicide" thesis, Turner argued for the relative impermanence of "civilization" in the face of the greater durability of human life. His ringing denunciation of cultural superiority surely would have alarmed white readers even as it reassured his black audience. He predicted, "The proud, haughty, domineering people of today may be the cringing, begging, sycophantic paupers of tomorrow." After all, he reasoned, "we have record of the rise to power and the decadence of various human tribes—black, yellow, and white," and while we "cannot know what characteristics or lines of conduct have the greatest survival value," he felt that "it should be the chief aim of courses in Human Biology to seek out and stress every factor which makes for peaceful, harmonious cooperation among races and among nations." Turner thus refocused the eugenic quest away from ensuring racial superiority, a quixotic venture requiring a static conception of fitness belied by life's dynamism. Turner's assimilationist eugenics freed him from the trap of biological superiority while acknowledging heredity's power.[27]

Turner taught assimilationist eugenics to thousands of students over more than 30 years. Yet, for these ideas to perform their antiracist, socially meliorative function, they had to be taught to those beyond the Talented

Tenth. "Average" African Americans (and ultimately whites, too) needed to internalize and act upon assimilationist eugenic principles. To spread the word, in 1927 Turner volunteered for the American Eugenics Society's list of eugenics lecturers. Years earlier, the NAACP baby contests provided another way to reach the public.[28]

BABY CONTESTS: EUGENICS AND FUND-RAISING

On March 20, 1925, a precocious 5-year-old girl closed the Newport News, Virginia, NAACP baby contest in verse:

> Good friends, we welcome you tonight
> To this our baby show,
> Just what we are trying here to do
> I'm sure that you all know.
>
> These babies all are hard at work
> For the National Association;
> If you will help and do not shirk,
> We'll have a better nation.
>
> Our mothers dear, as you can see,
> Have worked to win the prize,
> Just who will be the lucky one
> We can not now surmise.
>
> But whether the babies win or lose,
> Let us all stop and pause,
> And realize by buying votes
> We've helped a worthy cause.
>
> So once again we welcome you,
> With hearts all full of glee,
> Let us join in and give three cheers
> For the N.A.A.C.P.[29]

The poem outlined the contest's intended goals and hinted at its inspiration, guiding rationales, and purposes. Held between 1924 and 1934, and often overseen by NAACP Field Secretary William Pickens, these baby contests were designed with a twofold agenda. First, the NAACP sought to capitalize on a vogue in baby pageants and "Better Baby" and "Fitter Family" contests sweeping the majority white populace, adapting them to the black community and using them to raise funds for the NAACP's social justice campaigns. Second, these contests implicitly promulgated the assimilationist eugenics promoted by thinkers like Du Bois and Turner. Marrying eugenic theory with practical fund-raising, the NAACP's baby contests established a movement within the movement, one Du Bois dubbed the "Tenth Crusade," a sobriquet redolent with imagery of Christian salvation and noblesse oblige. The Tenth Crusade's ultimate goals of eugenic improvement might be understood as a class-biased attempt at intraracial social control. Nevertheless, the movement's proximate aim—raising money to fund the NAACP's battle against lynching—sought to improve life for all African Americans, regardless of class, birth, or "fitness."

EUGENIC CONTESTS ACROSS THE COLOR LINE

Scholars have traced the ways African American uplift ideology was shaped by, and in return shaped, the mainstream white eugenics movement. Yet, with the exception of Garveyite and white eugenic racial purity efforts, historians have overlooked programmatic similarities between black assimilationist and white mainline eugenics. Examining the NAACP's baby contests clarifies the structural affinities between black and white eugenics.

White eugenicists pioneered "Better Babies" and "Fitter Families for Future Firesides" competitions. Conducted at international expositions and state fairs across the country, these competitions—and their attendant eugenic exhibits—appeared at roughly 7 to 10 venues annually between the 1910s and the 1930s. Competitions peaked during the late 1920s, when as many as 40 potential sponsors annually solicited the Eugenics Record Office for guidance in hosting contests. Tens of thousands of white (and some black) attendees learned about eugenics and the goal of having a "goodly heritage" from these well-publicized competitions.[30]

Staged in the "human stock" sections of the fairgrounds, eugenics exhibits and contests explicitly analogized human and animal breeding. As Mary T. Watts, the co-organizer of the first Fitter Families Contest, said, "While the stock judges are testing the Holsteins, Jerseys, and whitefaces in the stock pavilion, we are judging the Joneses, Smiths, and the Johns." The stakes were high: contest brochures declared that the "science of human husbandry must be developed, based upon the principles now followed by scientific agriculture, if the better elements of our civilization are to dominate or even survive." Any "healthy" family could enter—they need only answer a lengthy questionnaire that elicited their "eugenic history." Judges—physicians, psychologists, social workers, and eugenics field workers—evaluated each family member for signs of superiority (special accomplishments and talents) and sought to ferret out mental, physical, and social defects. Individual scores were then compiled to arrive at a family's "grade." Final scores of B+ or above earned a bronze medal with "Yea, I have a goodly heritage" inscribed on its face. Organizers and eugenicists argued that the medalists' genetic fitness was "worth more than livestock sweepstakes or a Kansas oil well," because "health is wealth and a sound mind in a sound body is the most priceless of human possessions."[31]

This philosophy rang true for Du Bois, Turner, and many of their contemporaries among the black intelligentsia. They, too, concurred in the tautology that a healthy human race would be made up of healthy individuals. Witnessing the success white eugenicists achieved in mobilizing public interest, opinion, and money through eugenics contests, NAACP activists developed similar competitions: to address the "racial threats" confronting African Americans and to gin up support for assimilationist eugenics. These contests were the first systematic attempt to capitalize on eugenic interests in the black community.[32]

STRANGE FRUIT: HOW LYNCHING PROMPTED BABY CONTESTS

During the first third of the twentieth century, African Americans faced more immediate dangers than the dropping birthrate or "race suicide" that alarmed their white contemporaries. Between 1882 and 1944, American trees often bore "strange fruit": the bodies of some 4,700 African Americans, mostly men, often lynched by mobs, particularly in the South. Between 1909 and 1918, lynching claimed the lives of 687 people, of whom 590 were African Americans. Combined with race riots that rocked cities for days and the segregationist laws that whites insisted preserved peace by managing interracial contact, the lynching epidemic represented the nadir of American race relations.[33]

For Du Bois—whose motivation to organize what became the NAACP originated in a white storekeeper's display of a lynched man's knuckles—lynching represented the acme of inhumanity, an act that eviscerated white claims to superiority. Truly civilized people would never engage in such barbarism. Although whites claimed that lynchings only punished atavistic black "savages" who raped white women, beginning with Ida B. Wells's *The Red Record,* African Americans demonstrated that the vast majority of lynchings resulted from black/white male altercations and minor infractions of Jim Crow; barely a third stemmed from alleged interracial rapes. Lynching was most often the result of African Americans fighting oppression head on; from the perspective of Du Bois and many others, lynching thus claimed the very flower of black youth—the best and brightest who were unwilling to genuflect before racists.[34] Protest and political action embodied direct opposition to lynching; breeding an improved black population—one genetically predisposed to achieve beyond white expectation or reproach —offered another mode of resistance. Du Bois and other assimilationist eugenicists fought white supremacy on both fronts: the political and the biological.[35]

Du Bois railed against mob atrocities in the pages of the *Crisis,* alerting Americans to the scope and frequency of "The Lynching Industry." In 1922, the NAACP placed large ads in major white-owned newspapers presenting the facts about lynching, rending the threadbare white lie that murderous vigilantism was civilization's last resort against black devolution. That same year, Du Bois helped the "Colored Women of America" launch what he dubbed the "Ninth Crusade." Under the leadership of Mary B. Tolbert, this effort raised over $15,000 to combat lynching. But more could and would be done.[36]

THE TENTH CRUSADE: CONTEST LOGISTICS

Ever the pedant, Du Bois denominated NAACP anti-lynching efforts as the philosophical and numerical successors to what historians had long known as the "Eight Great Crusades" of Christendom. Success in the "Ninth" and "Tenth" crusades would exorcise lynching from America, bringing American civilization closer to the millennial peace promised in Christian theology. Recognizing the tremendous impact that eugenic ideology and the Fitter Family contests had on the white community, Du Bois hoped to adapt similar competitions to promote his agenda. He had already created something of a platform in the *Crisis's* "Children's Number." Meant to demonstrate the innate fitness of African Americans, the "Children's Number" also subtly inculcated assimilationist eugenics by valorizing youthful aspirants to the Talented Tenth as examples of sound black reproduction.[37]

Building on the success of the "Children's Number," Du Bois and William Pickens, the NAACP's national field secretary, launched a "Tenth Crusade." Eugenically fit African American infants and children would be the foot soldiers in this campaign. Having witnessed the success of baby contests as grassroots benefits for local chapters in Dayton, Cleveland, and Jersey City, Du Bois and Pickens believed that national contests could be cost-effective fund-raisers for the anti-lynching campaign. Presumably, in addition to mobilizing anti-lynching sentiment, identifying and voting for fit babies would help sensitize participants to the need for (and benefits of) rational eugenic procreation.[38]

The "nationalization" of the contests, and the merging of eugenic and financial priorities, complicated the straightforward logistics of the local competitions. At first, proud parents had paid an entrance fee to the sponsoring chapter to enroll their child in a contest; volunteer judges then selected the "best" baby. Winners were awarded gold coins; the chapter retained the proceeds. In July 1924, Pickens circulated a pamphlet outlining new, uniform competition guidelines.

In "How to Conduct a Baby Contest," Pickens created a hierarchy that privileged members of the Talented Tenth. Pickens stipulated that a sponsoring chapter must organize an executive committee—chaired by the chapter president's wife and composed of other influential members of the community—to oversee the competition. The executive committee would then appoint a publicity and meetings committee. Committee members and the chapter's officers importuned mothers to enter their fit babies in the contest; they simultaneously recruited the mothers as volunteer contest workers. Parent-volunteers were instructed to secure at least 10 additional workers for their child's "campaign"; each worker should raise $10 by selling votes for the child at a nickel apiece, NAACP memberships, and subscriptions to the *Crisis*. Theoretically, the "honor" of having one's baby selected would translate into a windfall for the association, as local chapters were instructed to send 75 percent of contest proceeds to the national office. These funds were then immediately channeled into the anti-lynching campaign. First, second, and third place winners from local contests proceeded to the national finals overseen by Pickens himself. For his part, Du Bois featured the finalists' photographs in the Children's Number, hoping for "all the good clear pictures of healthy human babies that we can get." Du Bois's emphasis on "healthy human babies" subtly reinforced assimilationist eugenics' central principle—that racial distinctions had no bearing on fitness.[39]

Rather than the "objective" evaluations of eugenic "fitness" that characterized white eugenic contests, the NAACP contests began by assuming fitness in the babies tapped to participate. Motivated by self-interest in crowning their children "fit" avatars of the race, parents (and their associates) were also overtly invested in funding the anti-lynching crusade. Personal politics, finance, and assimilationist eugenics thus worked together for the "greater good," rather than making the validation of rarified eugenic "fitness" and elitist social policy the contests' goal, as in white Fitter Families and Better Babies competitions. By March 1925, the "Brown Babies," as Pickens referred to them, had raised $12,000 for the anti-lynching cause.[40]

NAACP CONTEST RESULTS, CONTROVERSIES, AND LEGACY

Pickens was beginning to realize the full potential of what was before him. Local chapters were clamoring for additional recognition and publicity. Given "the money, pains, and loyalty" the chapters had thus far given the national headquarters, Pickens and Du Bois used the Crisis to reward local efforts and build momentum. By promoting the Cleveland chapter's contest of early 1925, for instance, Pickens hoped to "stimulate all pending contests," as well as "help him organize others." Du Bois featured the top three contestants, "along with the names of all chief contestants, brief notes of the contests, and the names of the officers of the organizing committees," in the "Weekly Press Stories" section of the Crisis. Such public recognition gratified grassroots activists, inspired additional contests, increased the overall financial yield, and popularized assimilationist eugenic precepts.[41]

This strategy worked well. By January 13, 1926, "the Colored Babies of the United States had raised more than $31,000.00 to fight for their own future." Pickens hoped that "every branch of the Association and every group interested in its work" would hold contests by early 1926. Additionally, he felt that "any branches that had held contests in previous years could and should repeat them," for "there are always new babies!" Pickens believed that contests represented "the least expensive and most successful way for local branches" to raise funds. By June 1926, 156 of the 300 local NAACP branches had held contests, raising another $10,000. This left "144 branches and a million babies" that should compete by the fall of 1926 or early 1927, according to Pickens. To spur action, Pickens mailed a letter claiming that organizing a competition was "simpler than riding a tricycle." Not afraid to tap into parental guilt, Pickens wrote that "any baby" would "enlist in the Crusade" against "color prejudice, unless some grown-ups stood in the way."

The Tenth Crusade was so successful that, by the 1930s, Pickens encouraged chapters to "expand their market," and consider hosting age-group contests for children ages 4–8, 9–12, and 13–15; "Young Women, Young Widows, Bachelors, Debutantes"; and even pastors. The possibilities seemed endless. Swept up in his own enthusiasm, Pickens apparently forgot that association members had finite personal finances.[42] Each new category of "eugenically fit" blackness offered the possibility of modeling eugenic ideals and raising revenue, but proliferating contests also began to raise tension along with funds. By exercising his editorial prerogatives through provocative opinion pieces and by promoting certain contest winners, with little or no input from other association officials, Du Bois often strained his relationship with the NAACP. Tensions came to a head in 1934, as Du Bois wrote a controversial series of editorials, "Segregation—A Symposium." His May editorial, in which he argued that "there should never be an opposition to segregation pure and simple unless that segregation does involve discrimination," seemed to announce a reversal of everything he and the NAACP had stood for over the preceding 25 years.

In actuality, Du Bois had simply returned to the arguments he had made 37 years earlier in "The Conservation of Races": that racial solidarity (what a later generation would refer to as "self-segregation" or "congregation in segregation") ultimately offered the power to "smash all race separation"—by allowing African Americans the space (environment) in which they might capitalize on the strengths of the Talented Tenth (heredity), uplifting the race (hereditarily and environmentally). With diplomatic understatement, and little appreciation of Du Bois's nuanced philosophical provocations, the NAACP board noted that Du Bois's position was not "in general agreement with that of the organization," and accepted Du Bois's resignation "with deepest regret" on July 10, 1934.[43]

With Du Bois's departure, the content and format of the *Crisis* changed. The assimilationist eugenic thrust disappeared with Du Bois. Coverage of the baby contests diminished significantly, even as the contests themselves dwindled and the general mood in the country drifted away from eugenics. Nevertheless, baby contests have persisted, sporadically, even down to today. Present NAACP contests no longer focus on fit babies. Instead, they promote the educational achievements of high school students, who compete for college scholarships on the basis of academic transcripts and essays. Even as the contests have lost their overt eugenic impulse, this "objective" evaluation of the contestants' "merits" echoes earlier white eugenic efforts to judge fitness by assaying intellect.

Analyzing the intellectual history of assimilationist eugenics and the organization of NAACP baby contests underscores at least four important implications about African Americans' flirtations with hereditarian ideology. First, members of the "mainstream" African American intelligentsia—thinkers like Du Bois, teachers like Turner, and their organization, the NAACP—used the ideological flexibility inherent in eugenic theory to advance their own political and social programs. Black eugenicists could and did "fight fire with fire" by inverting and adapting racist whites' eugenic ideas and turning them toward antiracist, equalitarian ends. Despite the obvious class bias inherent in the Talented Tenth, the goal of assimilationist eugenics was ever the improvement of the status of all African Americans and, ultimately, the *human* race.

Second, African Americans promoting assimilationist eugenics used the same fund-raising/proselytizing media as white mainline eugenicists: popular periodicals and baby contests. Du Bois embodied his hereditarian views in the Talented Tenth; he used the *Crisis* to promote his positive, assimilationist eugenic stance. Eugenic theory and political praxis met and married in the Tenth Crusade. The crusade capitalized on the eugenic vogue, dampened or deflected the harshest white charges about black hereditary inferiority, bolstered African Americans' self-esteem, and demonstrated the creativity of NAACP fund-raisers.

Third, while the Tenth Crusade baby contests might be viewed as cynical attempts by black elites to subject lower-class blacks to eugenic control, the tandem objectives of eugenic improvement and anti-lynching suggest a broader vision. The positive eugenic thrust—to increase the number in the Talented Tenth by educating people and using baby contests to set goals—contrasted with the negative eugenic emphasis underpinning the most successful white mainline eugenic programs: antimiscegenation, immigration restriction, and compulsory sterilization laws. By tapping into parental desires to promote and improve the quality of their children, the NAACP was able to raise over $80,000 (roughly $1 million in today's currency) between June 1924 and December 1930, four years before the contests ended.[44]

Finally, this study suggests that African Americans iterated what became the scientific consensus on race—that its hereditary determinants pale in significance compared to its environmental, cultural origins—some 30 years before white scientists acceded to it in the UNESCO Statement on Race, and almost 80 years before the Human Genome Project "proved" it. This prescience reveals the socially constructed nature of scientific understanding. Science is bound by the historical context in which it is conducted and influenced by the experience of the scientists conducting it. And these profound influences are antecedent to nonscientists' popularization of scientific findings. Ultimately, "good" or "true" science is as dependent on subjective criteria as it is on the purportedly "objective" findings elicited by the scientific method.[45]

With the diminution of the racism that separated white and black scientists at the beginning of the twentieth century, both sides converged on an understanding of heredity that moved beyond racial purity. The assimilationist eugenics of Du Bois, Turner, and the NAACP baby contests comported more closely with both the "facts" of genetics and the realities of a more integrated society—then and now. Assimilationist eugenics helped usher in today's more racially tolerant beliefs, politics, and science. That said, assimilationist eugenics raised the specter of continuing what

Elof Carlson has called, in his masterful understatement, "a bad idea": namely, the notion that some people are "unfit" and therefore less worthy of respect than others. Whether the legacy of assimilationist eugenics and the promise of genomics and genetic therapies will be realized in human harmony or invidious distinctions remains to be seen.

NOTES

1. *War against the Weak*, 35 mm, 93 min., Endless Films, New York, 2009; *Expelled: No Intelligence Allowed*, 35 mm, 97 min., Rocky Mountain Pictures, Salt Lake City, 2008.

2. See, for example, Mark H. Haller, *Eugenics: Hereditarian Attitudes in American Thought* (New Brunswick, N.J.: Rutgers University Press, 1962); Daniel J. Kevles, *In the Name of Eugenics: Genetics and the Uses of Human Heredity* (New York: Alfred A. Knopf, 1988); Nancy L. Gallagher, *Breeding Better Vermonters: The Eugenics Project in the Green Mountain State* (Hanover, N.H.: University Press of New England, 1999); Gregory Michael Dorr, *Segregation's Science: Eugenics and Society in Virginia* (Charlottesville: University of Virginia Press, 2008).

3. John David Smith, *Black Judas: William Hannibal Thomas and the American Negro* (Athens: University of Georgia Press, 2000); Martha F. Lee, *The Nation of Islam: An American Millenarian Movement* (Syracuse: Syracuse University Press, 1996); William A. Edwards, "Racial Purity in Black and White: The Case of Marcus Garvey and Earnest Cox," *Journal of Ethnic Studies* 15 (1987): 117–42. On Garvey and eugenics, see Michele Mitchell, *Righteous Propagation: African Americans and the Politics of Racial Destiny after Reconstruction* (Chapel Hill: University of North Carolina Press, 2004), 230–36.

4. This section builds on the work of Marouf A. Hasian Jr., *The Rhetoric of Eugenics in Anglo-American Thought* (Athens: University of Georgia Press, 1996), 51–71; Kevin K. Gaines, *Uplifting the Race: Black Leadership, Politics, and Culture in the Twentieth Century* (Chapel Hill: University of North Carolina Press, 1996); and Mitchell, *Righteous Propagation*, 86–88.

5. On "mainline" eugenics, see Kevles, *In the Name of Eugenics*, 88.

6. Susan Currell and Christina Cogdell, eds., *Popular Eugenics: National Efficiency and American Mass Culture in the 1930s* (Athens: Ohio University Press, 2006); Steven Selden, *Inheriting Shame: The Story of Eugenics and Racism in America* (New York: Teachers College Press, 1999), 39–62. On African Americans and popular eugenics, see Mitchell, *Righteous Propagation*, 75, 80–81, 86–88, 100–106. On fairs and contests, see Laura Lovett, *Conceiving the Future: Pronatalism, Reproduction, and the Family in the United States, 1890–1938* (Chapel Hill: University of North Carolina Press, 2007), 131–62; and Steven Selden, "Transforming Better Babies into Fitter Families: Archival Resources and the History of the American Eugenics Movement, 1908–1930," *Proceedings of the American Philosophical Society* 149 (June 2005): 199–225.

7. Gaines, *Uplifting the Race*, 80–83; Gregory Michael Dorr, "Beyond Racial Purity: African Americans and Integrationist Eugenics" (paper presented at the annual meeting of the Organization of American Historians, Memphis, Tenn., April 2003); Gregory Michael Dorr, "Fighting Fire with Fire: African American Intellectuals and

Hereditarian Thought, 1890–1942" (invited address, Wake Forest University Social Science Research Seminar, Wake Forest, N.C., October 2004).

8. On African Americans and nature/nurture (or eugenic/euthenics) distinctions, see Mitchell, *Righteous Propagation,* 147–49, 230; Du Bois, "The Talented Tenth," in *Writings by W.E.B. Du Bois in Non-periodical Literature Edited by Others,* ed. Herbert Aptheker (Millwood, N.Y.: Kraus-Thomson, 1982), 17–29; Du Bois, *Souls of Black Folk* (Millwood, N.Y.: Kraus-Thomson, 1989), 74–90, especially 87.

9. David Levering Lewis, *W.E.B. Du Bois: Biography of a Race, 1868–1919* (New York: Henry Holt, 1994), 170–74. Quotations from Du Bois, "The Conservation of Races," in *Pamphlets and Leaflets,* ed. Herbert Aptheker (White Plains, N.Y.: Kraus-Thomson, 1986), 2–3, 4, 5.

10. Du Bois, "Conservation of Races," 6; Du Bois, *Souls,* 9.

11. Du Bois, "The Talented Tenth," 17–18.

12. Du Bois, "Heredity and the Public Schools," in *Pamphlets and Leaflets,* 50, 51.

13. On the "submerged tenth," see English, *Unnatural Selections,* 45. For dysgenic whites, see Paul A. Lombardo, *Three Generations, No Imbeciles: Eugenics, the Supreme Court, and Buck v. Bell* (Baltimore: Johns Hopkins University Press, 2008), 134; Du Bois, "Marrying of Black Folk," *Independent* 69 (October 13, 1910): quotations 812, 813; English, *Unnatural Selections,* 44.

14. On relations between the "best stocks," Du Bois wrote, "Here I suddenly saw that not only in some far future world our Talented Tenth with the talented of all nations leads all people to salvation." *Non-periodical Literature,* 196; see also "Sociology and Industry in Southern Education," in *Writings by W.E.B. Du Bois in Periodicals Edited by Others,* 4 vols., ed. Herbert Aptheker (Millwood, N.Y.: Kraus-Thomson, 1982), 371; Du Bois, *Souls,* 48, 71–72, 136.

15. Du Bois, *Souls,* 70; Du Bois, "Editorial," *Crisis* 4 (October 1912): 287; Du Bois, "The Immediate Program of the American Negro," *Crisis* 9 (April 1915): 311.

16. Du Bois, "Opinion," *Crisis* 24 (August 1922): 152–53; Du Bois, "Opinion," *Crisis* 24 (October 1922): 199.

17. English, *Unnatural Selections,* 48; Albert Sidney Beckham, "Applied Eugenics," *Crisis* 26 (1924): 177; Hasian, *Rhetoric of Eugenics,* 64–69.

18. English, *Unnatural Selections,* 55, 53.

19. Ibid., 49. "Five Exceptional Negro Children," *Crisis* 34 (October 1927): 258.

20. Hasian, *Rhetoric of Eugenics,* 69–70; W.E.B. Du Bois, "Black Folk and Birth Control," *Birth Control Review* 16 (1932): 166, 167; Du Bois, "Opinion," *Crisis* 24 (August 1922): 153.

21. Dorr, *Segregation's Science,* 98–104.

22. Ibid., 101. The exams are in the Thomas Wyatt Turner Papers, University Archives, Hampton University. Turner probably also taught eugenics at Tuskegee, but documentary evidence of this is lacking.

23. On uplift, see Gaines, *Uplifting the Race,* xiv–xv; Du Bois, *Souls,* 79; English, *Unnatural Selections,* 35–36. See also Daylanne K. English, "W.E.B. Du Bois's Family Crisis," *American Literature* 72 (2000): 292. For eugenics and club women, see "Child Welfare and Mat[e]rnity" (1928), in *Records of the National Association of Colored Women's Clubs, 1895–1992,* ed. Lillian Serece Williams (Bethesda, Md., 1993), reel 7, frames 00355–58.

24. Thomas Wyatt Turner, "Feeblemindedness" and "Lecture X—Eugenics," Turner Papers.

25. Thomas Wyatt Turner, "The Biological Laboratory and Human Welfare," *Howard University Record*, January 1924, 4, 5.

26. Ibid., 7.

27. Thomas W. Turner, "The Curriculum and Aims in Biological Teaching," *School Science and Mathematics* 27 (October 1927): 689–90.

28. American Eugenics Society, "A List of Eugenics Lecturers (1927)," 14, in "AES Printing Orders, 1926–1942" folder, American Eugenics Society Records, American Philosophical Society Library, Philadelphia.

29. William H. Pickens to James Weldon Johnson, March 10, 1925, Records of the NAACP, "Branch Files: 1910–1947," Part I, box I:G208, Records of the National Association for the Advancement of Colored People, Manuscript Division, Library of Congress, Washington, D.C.

30. Lovett, *Conceiving the Future*, 131–62; Selden, *Inheriting Shame*, 22–38.

31. Quotations from Lovett, *Conceiving the Future*, 144; Kevles, *In the Name of Eugenics*, 62. Individuals with averages of B+ but with even one subscore lower than B did not win medals. Selden, *Inheriting Shame*, 32–33 (and image of medal).

32. On earlier beauty contests and black popular eugenics, see Mitchell, *Righteous Propagation*, 213–14.

33. English, *Unnatural Selections*, 119. The literature on lynching is massive; estimates of numbers vary based on the definition of what constitutes a lynching. See Grace Elizabeth Hale, *Making Whiteness: The Culture of Segregation in the South, 1890–1940* (New York: Pantheon Books, 1998), chapter 5 and 355n4. For statistics, see Arthur F. Raper, *The Tragedy of Lynching* (New York: Dover, 1970), 480–84.

34. English, *Unnatural Selections*, 27.

35. Hale, *Making Whiteness*, 214–15; Lewis, *Du Bois: Biography*, 226; Ida B. Wells-Barnett, *A Red Record: Tabulated Statistics and Alleged Causes of Lynching in the United States, 1892, 1893, 1894* (Chicago: Donohue and Henneberry, 1895).

36. Robert L. Zangrando, *The NAACP Crusade against Lynching, 1909–1950* (Philadelphia: Temple University Press, 1980); Patricia A. Schecter, "Unsettled Business: Ida B. Wells against Lynching; or, How Lynching Got Its Gender," in *Under Sentence of Death: Lynching in the South*, ed. W. Fitzhugh Brundage (Chapel Hill: University of North Carolina Press, 1997), 309.

37. See NAACP, "Crusade 1922" and "Crusade 1923" folders in "Administrative Files: 1885–1949," Part I, boxes I:C206 and I:C207, Records of the National Association for the Advancement of Colored People, Manuscript Division, Library of Congress, Washington, D.C.; Du Bois, "The Ninth Crusade," *Crisis* 25 (March 1923): 214. The Children's Number ran every October.

38. See "Benefits—Baby Contests 1924–1932" folders in box I:C215, NAACP Administrative Files. Hereafter baby contest files.

39. William Pickens, "How to Conduct a Baby Contest," baby contest files; Du Bois, "Opinion," *Crisis* 26 (September 1924): 199. For examples of the photos, see English, *Unnatural Selections*, 50–51.

40. Baby contest files.

41. Ibid.

42. Ibid.

43. Du Bois, "Segregation," *Crisis* 36 (May 1934), quoted in David Levering Lewis, *W.E.B. Du Bois: The Fight for Equality and the American Century, 1919–1963* (New York:

Henry Holt), 336; Du Bois, "Counsels of Despair," *Crisis* 36 (June 1934), quoted in ibid., 345. Du Bois's resignation and the board's acceptance appear in *Selections from the* Crisis, vol. 2, *1926–1934*, ed. Herbert Aptheker (Millwood, N.Y.: Kraus-Thomson, 1983), 770–72. This paragraph accepts and extends Lewis's interpretations of this pivotal struggle between Du Bois and the organization he founded.

44. Baby contest files. Total fund-raising figures could not be calculated.

45. Michelle Brattain, "Race, Racism, and Anti-Racism: UNESCO and the Politics of Presenting Race to the Postwar Public," *American Historical Review* 112 (December 2007): 1386–1413.

PART THREE

State Studies of
Eugenic Sterilization

FIVE

From Legislation to Lived Experience: Eugenic Sterilization in California and Indiana, 1907–79

ALEXANDRA MINNA STERN

In February 1950, the Fort Wayne State School in Indiana held a steriliza-tion hearing for Vernon, a young biracial man. Since his original commit-ment a decade earlier, at the age of 9, Vernon had managed multiple es-capes from the institution, and it was during one of his recent flights that the superintendent decided it was necessary to sterilize Vernon "for the protection of the community."[1] Following the requirements of Indiana's sterilization law, the superintendent filed a petition with the Fort Wayne State School's Board of Supervisors, asking them to approve Vernon's vasectomy based on the medical judgment that he was "definitely feeble-minded and incurable" and that his welfare and that of society would be promoted by his sterilization. In addition, notices about Vernon's steril-ization hearing were served on him and his parents and sent to Indiana's Public Welfare Department. Assuming the surgery occurred 30 days after the hearing, which was the protocol, Vernon was one of 84 inmates steril-ized at Fort Wayne in 1950.

One month earlier and 2,000 miles away, the sterilization of a 14-year-old Latina at the Pacific Colony in Spadra, California, was approved by the director of the Department of Mental Hygiene. In compliance with the state's legal provisions, Pacific Colony's superintendent checked the boxes on the sterilization order form indicating that the operation was warranted because Esperanza, who had spent three years at the institu-tion, was afflicted with "feeble-mindedness" of the "imbecile grade" and "perversions or marked departures from normal mentality."[2] Following departmental procedures, written consent for Esperanza's operation had been obtained from her father. Given that her sterilization request was processed smoothly and with no objections, Esperanza was almost certainly one of the 45 patients, 30 of whom were female, who were sterilized at Pacific Colony in 1950.[3]

These two snapshots of institutionalization and sterilization capture vital aspects of the intersecting histories of eugenics, mental health, disability, and the state regulation of reproduction in twentieth-century America. Like many of the more than 60,000 patients and inmates sterilized in institutions in 33 states from the early 1900s to the 1970s, Vernon and Esperanza were teenagers with unstable family situations whose initial encounters with the county or juvenile court eventually resulted in commitment to a feebleminded home. Their sterilizations were ordered based on their mental classification as imbeciles and their alleged destructive personalities, associated with thievery, truancy, and petty crime in Vernon's case, and "social" and "glandular" problems—euphemisms for perceived promiscuity and sexual impropriety—in Esperanza's case. More broadly, these two examples highlight themes, such as the interplay of race, gender, and class, as well as the formidable power of medical superintendents, clinical diagnosis, and bureaucratic habituation, that scholars have examined over the past few decades in a fruitful effort to broaden knowledge about the rise and fall of eugenic sterilization in modern America.[4]

Yet the stories of Esperanza and Vernon also point to a series of unexplored questions. These surgeries were ordered in 1950, at the twilight of the era of state sterilizations. Within the next five years, operations in California and Indiana would drop considerably due to overlapping legal, administrative, and attitudinal changes. Given that studies of sterilization and eugenics overwhelmingly concentrate on the period before World War II (even though many states maintained robust sterilization programs even into the 1960s), it is important to explain why sterilizations did not abate until the mid-1950s and what impelled their conspicuous decline. Explaining the reasons for this intriguing shift is particularly instructive for Indiana and California, which shared the distinction of spearheading eugenic sterilization even as their laws diverged markedly in terms of acceptance and implementation.

These two stories also are illuminating because they offer glimpses into the lives of those sterilized and the social worlds of institutions where thousands of children and adults were committed because of their presumed mental defects and behavioral problems. Just as we know very little about why approximately 15 states never passed sterilization laws in the twentieth century, puzzling questions remain about why institutions varied so widely from state to state and in the same state. Some homes and hospitals carried out hundreds, even thousands, of procedures while others carried out few, if any. In a related vein, because patient records are so difficult to find, let alone access, the people whose lives were irreparably altered by institutional sterilizations are often absent in historical studies.

Situating the experiences of Esperanza and Vernon in the wider context of eugenic sterilization in Indiana and California can shed light on lived experiences of reproductive surgery, institutional differences, the demographics of patient populations, the role of state authorities and family members in sterilization proceedings, and patterns of change over time. As windows onto the politics and practices of sterilization in the mid-twentieth century, these stories help to illustrate the elasticity and longevity of eugenic practices and ideas in two states that played critical roles in American eugenics.

CONVERGING AND DIVERGING STERILIZATION LAWS

Indiana and California introduced eugenic sterilization statutes to America, passing the first and third laws, respectively, in 1907 and 1909 (Washington passed a statute a few weeks before California).[5] Indiana's sterilization law was in many respects a legislative afterthought formulated to shield Dr. Harry Sharp, the medical superintendent at the Indiana Reformatory in Jeffersonville, who had started to perform vasectomies on prisoners in 1899. Initially, Sharp's justification for these procedures was therapeutic, specifically to alleviate imprisoned men of excessive masturbation.[6] However, over the years, as he honed his surgical technique, Sharp began to see sterilization not just as a potential cure for supposed sexual disorders but as a method to prevent the transmission of the bad heredity of criminals from one generation to the next. As he wrote in the pamphlet *Vasectomy: A Means of Preventing Defective Procreation,* "restricting propagation seems to be universally agreed on as necessary for the relief of the downward tendency."[7]

In March 1907, Sharp and the general superintendent of the Indiana Reformatory, William H. Whittaker, collaborated with a state representative to introduce an act to "prevent the procreation of confirmed criminals, idiots, imbeciles, and rapists."[8] Urging the passage of this legislation, Sharp wrote, "No confirmed criminal or other degenerate ever begot a normal child, and for this reason I enter the plea for society in general and for the unborn child in particular that this bill be enacted into law."[9] Sharp believed this act simultaneously would save the state thousands of dollars by allowing for the release of treated inmates and halt the transmission of "mental as well as physical defects" to offspring.[10] After the governor signed this bill into law in April 1907, Sharp wasted no time, sterilizing 119 men in 1908 (over 10 percent of the reformatory's average patient population of 1,100).[11] In 1909 he reported that he had sterilized 456 men since 1899 and expressed his impatience to implement the "Indiana Plan" in facilities beyond Jeffersonville.

Despite Sharp's enthusiasm, Indiana's governor, Thomas R. Marshall, was wary of Sharp's activities and unsure of the statute's constitutionality, and in spring 1909 he ordered a moratorium on sterilizations.[12] Subsequent governors upheld this de facto ban on sterilizations, a decision in keeping with the contested status of sterilization laws in the 1910s, as acts passed in other states were challenged on various grounds and often rendered unconstitutional in the courts. In Indiana, politicians were particularly reluctant to countenance an act with virtually no legal protections for patients or inmates.[13] Seeking some resolution, in 1919 Governor James Goodrich decided to test the law's constitutionality. He appointed the Jeffersonville city attorney to defend Warren Wallace Smith, convicted of rape and incest, against a sterilization order approved by the Indiana Reformatory's board of trustees.[14] After a decision for Smith in the Clark Circuit Court, Dr. Charles F. Williams, the reformatory's chief physician, appealed to the Indiana Supreme Court, which in turn sustained the lower court's decision in 1921. The Indiana Supreme Court clarified that the sterilization law violated the state constitution and the U.S. Constitution, specifically the Fourteenth Amendment, by depriving Smith "of life, liberty and property without due process of law" as well as the "equal protection of the laws." This decision also stated "while vasectomy is physically less severe than castration, in its results it is much the coarser and more vulgar, and is equally cruel and inhuman."[15]

This unambiguous interpretation laid the groundwork and set the limits for Indiana's 1927 act, which pertained exclusively to institutions for the feebleminded, insane, and epileptic, and was written to stress the preventive health benefits of insulating the populace against defective heredity. After the passage of this new act, Indiana carried out the majority of its sterilizations, about 2,000 (of approximately 2,500 total from 1907 to 1974). Over 1,500 or 75 percent of these were performed at the Fort Wayne State School, which had been founded as an orphans' home in 1879 and designated a facility for feebleminded children in 1890. This institution housed patients of all ages, who were instructed in basic subjects and taught gender-specific skills like cooking and table waiting (girls) and shoemaking and brush making (boys) to be able to work in the hospital and following possible out-placement and parole.[16] In 1920, excess capacity prompted the state to establish the Muscatatuck State School, which claimed the remaining 500 surgeries.[17] From 1927 until the early 1950s, the superintendents and resident physicians at both of these Indiana institutions saw sterilization as a "progressive step in the state's mental hygiene program," and in 1932 they made it a precondition for release.[18]

California's law did not travel along such a disrupted path. Although revised, amended, and even substantially modified, the Golden State's sterilization statute remained on the books without interruption for 70 years. Envisioned by F. W. Hatch, the secretary of the State Commission in Lunacy [sic], this act was sponsored by Senator W. F. Price of Santa Rosa in February 1909. It passed the Senate with only one nay vote, passed the House unanimously, and was signed into law by the governor in April 1909. It granted superintendents the authority to "asexualize" a patient or inmate if such action would improve his or her "physical, mental, or moral condition."[19]

Although it initially targeted the feebleminded, in 1913 California's sterilization law was broadened to apply to an assortment of inmates and patients, including those in the Sonoma State Home, insane hospitals, and prisons (with the qualification of having been convicted of two sex crimes or three other crimes and identified as a "sexual pervert"), as well as idiots and those "afflicted with hereditary insanity or incurable chronic mania or dementia." This modification also unconditionally immunized any official participating in sterilization procedures, regardless of patient consent, against civil and criminal liability. Four years later legislators revised the law again, adding more overt eugenic language. Now the act pertained to those "afflicted with mental disease which may have been inherited and is likely to be transmitted to descendents, the various degrees of feeblemindedness, those suffering from perversion or marked departures from normal mentality or from disease of a syphilitic nature."[20] The 1917 amendment also made sterilization a precondition for institutional release and incorporated the newly approved Pacific Colony, where Esperanza would be sterilized 33 years later, to the list of affected institutions. Finally, this revision stipulated that the written consent of, or a written request from, the parent or guardian was needed to operate on "any idiot, if a minor."[21] In other words, according to the law, only in the most extreme circumstances—of a child with a presumed mental age under three—was consent nonnegotiable.

Performed sporadically at first, the number of sterilizations in California gradually climbed while the legislation was being revised and expanded. By 1921, California accounted for 2,248 sterilizations—over 80 percent of all cases nationwide.[22] This trend continued into the 1920s, as an extensive eugenics network consolidated in the state, and accelerated in the 1930s.[23] The pervasive acceptance of a preventive health rationale for reproductive surgery, the pressures of institutional overcrowding, and the fiscal constraints caused by the Great Depression led to a considerable rise in sterilizations in California in the 1930s. Like in other states, sterilization was embraced by superintendents as a remedy for excess commitments, by psychiatrists as a prophylactic measure needed to release patients into supervised settings, by some inmates and patients as a viable ticket out of the institution, and by some relatives as a credible method of normalization and behavioral control. The country as a whole saw a marked rise in sterilizations during this period, fueled in part by the 1927 landmark decision *Buck v. Bell,* in which the U.S. Supreme Court upheld Virginia's sterilization statute.[24] By 1932, 27 states had laws on the books, and the annual number of operations nationwide peaked at just over 3,900.[25] Mimicking these trends, operations in California climbed from several per year in the early 1920s to over 800 by the 1930s; by 1950 over 18,000 sterilizations had been performed in the Golden State.[26] In the end, more than 20,000 surgeries, or one-third of all those nationwide, were carried out in California institutions, easily making the state the leader in overall sterilization statistics.

Despite its resemblance to comparable laws around the country, California's sterilization act stands out for its omissions. For over four decades, it provided no legal mechanism for an inmate or patient to challenge a sterilization order, required no written notification to the patient and to the parent or guardian except in the case of minor "idiots," and offered no opportunity for a hearing at the institutional level. Probably cognizant of the statutes that had been overturned in states including Iowa, Michigan, and Indiana, and eager to afford some legal protection to state agencies and authorities, in the early 1920s the Department of Institutions mandated that all superintendents send signed sterilization requests to Sacramento for the director's approval, preferably accompanied by the written consent of a parent or adult guardian.[27] The fact that a high percentage of the approximately 15,000 sterilizations received by Sacramento between 1922 and 1952 included signed consent forms indicates that authorities took this directive very seriously even though it was not an obligation under the law.

Furthermore, California was exceptional in that its sterilization law never faced any serious legal challenge.[28] Indeed, only one case, *Sara Rosas Garcia v. State Department of Institutions,* rose to the appellate level. In 1939 when this suit was filed, Sara's daughter Andrea was a patient at the Pacific Colony, having become a ward of the Los Angeles County Juvenile Court one year earlier. Sara was a widow with nine children who ranged in age from Andrea, 19, to Ricardo, 2. Notwithstanding the pressures of raising so many children, Sara acquired pro bono legal counsel and filed a Writ of Prohibition to prevent the Pacific Colony superintendent from performing a salpingectomy on her eldest daughter. Sara's attorney argued that California's sterilization law violated the Fourteenth Amendment of the federal constitution and the equal protection and due process clauses of the state constitution. He added that the surgery would be performed against Andrea's "wishes and desires" and that the law gave "no remedy or method of redress" for the "irreparable damage" she would suffer. Although Sara's writ was denied in a 2 to 1 decision, Judge J. White lambasted the existing law in a terse dissent. In White's opinion, a sterilization order, insofar as it deprived a person of the "right of procreation," was consequential enough to merit judicial consideration beyond the purely administrative arena of the Department of Institutions. White wrote that "the grant of such power should be accompanied by requirements of notice and hearing at which the inmate might be afforded an opportunity to defend against the proposed operation. To clothe legislative agencies with this plenary power, withholding as it does any opportunity for a hearing or any opportunity for recourse to the courts, to my mind partakes of the essence of slavery and outrages constitutional guaranties."[29] Despite her legal resistance, records indicate that ultimately Andrea was sterilized at Pacific Colony in 1941.

Indeed, requirements mentioned by White were not added until the early 1950s, when successful Senate (1951) and Assembly (1953) bills deleted any references to syphilis (long since understood as bacterial and not hereditary in etiology) and sexual perversion, removed references to the feebleminded, "idiots," and "fools," terms seen as archaic, from the purview of the law, and instated patient and next-of-kin notification as well as channels for legal appeal at the county court level.[30] This substantial revision was supported by the Roman Catholic Bishops of California, who opposed the existing law in its entirety, and the Department of Mental Hygiene (previously named the Department of Institutions), which wanted to modernize the nomenclature used to describe the mentally ill and retarded.[31] Approved by Governor Earl Warren, these significant amendments also reflected his administration's commitment to reviewing and updating California's mental health and public health programs.[32]

These changes had a dramatic impact. What had once been a mere formality became a much more taxing procedure, deterring many physicians from seeking sterilization orders.[33] From 255 operations in 1951, the number dropped to 51 in 1952, and by the mid-1950s it hovered below 20, even as three new hospitals and homes opened.[34] For example, at the Sonoma State Home, which by 1950 had carried out about 5,500 sterilizations, only 4 were performed in fiscal year 1952–53 and 1 in 1953–54. The superintendent at the time noted the "conspicuous drop" in the number of surgeries, which he attributed to a variety of factors including the modified law.[35] The glaring lack of institutional oversight or legal recourse for patients from 1909 to 1951 helps to explain why, even when compared per capita to states with much smaller populations, sterilization rates in California were always high.[36]

Comparing Indiana and California, forerunners of eugenic legislation and activism, demonstrates that sterilization laws in the United States were neither monolithic nor unified. For instance, Hoosier legislators and superintendents exhibited greater awareness than their West Coast counterparts of the need for layered procedures that could offer patients some legal and administrative latitude. Indiana's lengthy 1927 act delineated a two-stage procedure by which the superintendent presented a petition to the institutional governing board, scheduled a hearing to validate the request, and served copies of the petition on the inmate and next of kin with at least 30 days anticipation, followed by a board hearing with the inmate and sometimes a relative present for the official approbation. Bracing themselves against the weaknesses of the 1907 and analogous acts overturned in state courts, the crafters of this legislation inserted sections that allowed for a right to appeal the decision to the circuit court (and in the next instance, the Indiana Supreme Court) and clarified that this law could not be construed to permit castration or organ removal. Yet Indiana legislators also were keen to shield themselves from potential litigation and establish state authorities as the final arbiters of sterilization decisions. Emulating California, this act immunized officials from civil or criminal liability and granted the institutional board the power to order, despite dissent, the sterilization of any inmate found "by the laws of heredity" to be a "probably potential parent of socially inadequate offspring."[37]

In practice, Indiana's 1927 law translated into board hearings where sequential sterilization orders were issued, usually based on a feebleminded diagnosis, and required certification from a physician or psychologist that an inmate or patient would be the "probably potential parent of socially inadequate offspring" and that his or her welfare and that "of society will be promoted by such sterilization."[38] However, unlike California, where sterilizations were decided solely in the institutional arena, in Indiana a 1931 amendment conferred this authority to the county court, empowering the judge and two testifying physicians to authorize sterilization at the inquests of the feebleminded, the insane (per a 1935 amendment), and the epileptic (per a 1937 amendment). Thus Indiana's sterilization program was uniquely two-pronged in that it enabled both county courts and institutional boards to order sterilization. According to Dr. L. Potter Harshman, Fort Wayne's psychiatrist and a regular presence at board hearings, flexibility characterized this dual system. Speaking before an audience of his colleagues in the American Society for Mental Deficiency, Harshman admitted that "perhaps this convenient arrangement has the proportions of being a little too wholesale in the minds of most of you," but he justified the policy as "progressive" because it enabled more defectives to be transitioned into home care.[39]

Once its dual system was in place, Indiana, which ranked seventh nationwide in total sterilizations, registered a fairly steady rate of operations from the 1930s to the 1950s. These peaked at 159 in fiscal year 1945–46 and only began to fall significantly in fiscal year 1955–56, to 36, then to 24 in 1956–57, to 15 in 1959–60, and finally to 2 in 1961–62.[40] In contrast to California, what prompted this decrease was not legislative revision but the 1956 reorganization of the Department of Mental Health.

In general, the plummeting sterilizations in both Indiana and California revealed a metamorphosis in approaches to mental retardation. As a Sonoma State Hospital physician wrote in 1956, mental retardation was "as much a social and psychological problem as it is a medical one."[41] No longer a hereditary stigma to be eradicated, mental retardation was increasingly viewed as a condition that should be discussed and treated openly among doctors, parents, and the retarded themselves. The founding of the National Association of Retarded Children in 1950 and the increasing devotion of resources to mental health on the federal level that followed the 1946 creation of the National Institute of Mental Health symbolized this sea change.[42]

INSTITUTIONAL LIVES

One month after she was committed to California's Patton State Hospital in February 1939, Rhonda, a white 26-year-old mother of three children, was sterilized.[43] According to her sterilization order, Rhonda qualified for salpingectomy under the law on three counts: she was schizophrenic and catatonic, infected with gonorrhea, and categorized as an imbecile with an IQ of 46. Given that Rhonda's personal history noted that she "got drunk, left home," and "associated with other men," it is not surprising that her husband provided written consent for the operation just three days after her commitment.[44]

Rhonda's operation took place at the apex of California's sterilizations, fiscal year 1939, when 828 procedures were performed in nine institutions. Patton accounted for 180 of these, 98 on men, and 82 on women.[45] Located in San Bernardino and established as the main mental hospital for southern California, Patton opened to its designated patients—the insane, inebriates, and drug addicts—in 1893. Since the advent of the sterilization era in 1909, Patton's superintendents had been vocal advocates of reproductive surgery for therapeutic and eugenic purposes. For example, in 1916, superintendent Dr. John A. Reily wrote in response to a survey sponsored by the California Department of Charities and Corrections that the aim of sterilization was to improve "the standard of the human race," adding that the occasional denial of the pleasures of parenthood was a "small consideration as compared with the vast benefits accruing to society in the prevention of the propagation of the unfit."[46] Ten years later, his successor, Dr. G. M. Webster, described the centrality of sterilization to Patton's treatment methods: "We are trying, in so far as possible, to sterilize every male and female who enters the hospital during active sexual life," not only to relieve the inmate's "present mental condition" and avert future attacks but also to limit "as far as possible the birth of the unfit into the world."[47]

The zeal for the "surgical solution" at Patton, which carried out more than 4,500 operations between 1909 and 1950, helped to ensure that California's mental hospitals performed more total sterilizations (approximately 12,000) than its feebleminded homes (approximately 8,000).[48] To a great extent, more operations occurred in mental hospitals because California maintained seven such facilities (Patton, Agnews, Norwalk, Stockton, Camarillo, Mendocino, and Napa) as opposed to two feebleminded homes (Sonoma and Pacific Colony). In terms of sheer population size, the mental hospitals housed an average of five times as many inmates than the feebleminded homes. For example, in 1940, California's mental hospitals held 22,953 compared with 4,076 in Sonoma and Pacific Colony combined.[49]

All of California's mental hospitals were crowded; for example, between 1910 and 1955 the total resident population increased more than fivefold, from 6,864 to 36,403. Yet Patton was consistently overflowing. In absolute terms, its population grew from 1,372 in 1910 to 4,128 in 1950.[50] In the 1930s, Patton regularly reported an excess of at least 50 percent. In 1939, Patton held 3,843 inmates despite a certified capacity of 2,983.[51] In contrast, Stockton State Hospital had a lower percentage of excess inmates despite a higher patient capacity and population. Even though its superintendent, Dr. Margaret Smyth, was a tireless proponent of the

eugenic virtues of sterilization, Stockton always reported fewer operations than its sister institution.[52] Since sterilization was a precondition for
release, it is very likely that Patton's severe overcrowding played a key role
in its elevated sterilization rates and probably contributed to the brief wait
between commitment and sterilization experienced by Rhonda.

In marked distinction to California, only a handful of sterilizations
took place at mental hospitals in Indiana. Given that Indiana maintained
six such facilities (Logansport, Indiana Hospital for Insane Criminals,
Central State, Evansville, Madison, and Richmond) but only two feebleminded homes (Fort Wayne and Muscatatuck), the state's overwhelming
concentration on the feebleminded is curious. If on paper both states'
sterilization laws applied equally to the insane and the feebleminded,
what accounts for this discrepancy? A partial answer is timing. In California, sterilizations began to accelerate in the early 1920s, just as Indiana's
1907 law was ruled unconstitutional and about a decade before Indiana
had fully established its two-pronged system of board- and court-ordered
sterilizations. Hence, California had a 10-year lead on Indiana, which
coincided with the decade before many psychiatrists started to lose faith
in the value of sterilization and turn to new treatment modalities such as
electroshock and insulin therapies. As this shift occurred, California's
feebleminded homes started to outpace the state's mental hospitals in
sterilization rates, a transition that was complete by the mid-1930s and
involved a gender reversal as more operations began to be performed on
women than men.[53]

From today's vantage point, many of those sterilized, whether in
mental hospitals or feebleminded homes, would be considered mentally
or developmentally disabled. In retrospect, it is also clear that many were
young women and men who ended up in state facilities because of the
social and psychological impact of poverty, lack of education, broken
homes, petty crimes, and societal anxieties over "sexual immorality"—
which for girls might mean sexual activity (whether consensual or forced)
and for boys same-sex intimacy. Beyond fulfilling the eugenic aim of
denying those labeled defective the ability to procreate, sterilization frequently functioned as a form of behavioral management, employed by
superintendents who wanted to discharge disruptive inmates, or relatives
who believed that a loss of reproductive capacity could tame their son,
daughter, or wife. Finally, even though operations were performed based

on laws that avoided the language of punishment or physical retribution, frequently there was a punitive dimension to eugenic sterilization, particularly for patients or inmates classified as sexually deviant.

Patient records reveal that parents and spouses often were very involved in sterilization decisions. Rhonda's husband, for example, swiftly consented to her sterilization. In the case of Daphne, another patient at Sonoma, who became a ward of the court at 16 when it was discovered that she was four months pregnant and infected with gonorrhea, her parents requested sterilization upon her commitment in 1928. As her sterilization order explained, Daphne "has recently given her parents renewed trouble due to her conduct with men and boys; therefore they request her sterilization."[54] In 1932, the mother of 15-year-old Oliver, also at Sonoma, asked Butler to perform a vasectomy on her son, classified as a middle moron with an IQ of 61, because he was "rough with other children and a menace to the neighborhood when on parole."[55] Similar patterns were at play in Indiana. In 1942, the parents of 19-year-old Camille readily agreed to their daughter's sterilization. Described as "rather attractive" and "well-developed physically," Camille was admitted to Muscatatuck at the age of 16 because she had naively sought out the affections of men and contracted venereal disease. Three days after a social worker made a "sterilization visit" to the home of Camille's parents and told them that their daughter, whom they missed terribly, would be furloughed only after a salpingectomy, they sent a note to Muscatatuck asserting that they were willing to "have their darling sterilized." Several months after the required paperwork was processed, and a board hearing held at which the parent's note was read, Camille's surgery was realized at the Indiana University Medical Center.[56]

Some parents vehemently objected to institutional requests to sterilize their children. For example, the same year that Camille's parents assented to their daughter's sterilization, the mother and aunt of a female minor at Indiana's Fort Wayne State School went before the board, with a letter in hand from a physician stating that the proposed sterilization should not be performed. Nonetheless, the mother and aunt were informed that the operation could only be stopped with expert certification that their kin was not feebleminded, a document they apparently never obtained. After an internal discussion, the board concluded that the physician's letter was not sufficient. Claiming that it had made its decision reluctantly, and iterating that the two petitioners could file an appeal if desired, the board contended that it was merely following the law, which had been crafted with "the thought of protection of the inmate as well as society."[57] Eleven years later, the board heard from another patient's mother, who testified that her late husband had professed his resistance to the sterilization of their institutionalized son, an opinion with which she concurred. She told the board, "Because I feel when he [her son] is released from this place, he will want to get married and he and his wife would like to raise a family." When the board asked the boy, also present at the hearing, if he would agree to a vasectomy, he stated plainly, "I don't want it." Given this opposition, the board placed this decision in abeyance, and extant records indicate that this patient was never sterilized.[58]

The success in thwarting sterilization requests that a small number of relatives in Indiana achieved was much more elusive in California. Although superintendents almost always sought written consents from the parents or adult guardians of their charges, this interaction ended on a two-dimensional piece of paper. With no board hearings to attend and legal avenues closed, relatives could do very little to voice their dissent. For example, in November 1931, Butler wrote to Mr. Romero, the father of Juan, a minor at the Preston School of Industry, asking him to consent to his son's sterilization at Sonoma so "he can never reproduce his kind, for we know from experience that individuals of his mentality should never bear off-spring, as they are usually defective in some manner."[59] One of Mr. Romero's three sons, Javier, had already been sterilized, and in the same communication, Butler reminded him that Sonoma was still awaiting approval for his third son, Pablo. Butler asserted that having three boys in one family who ended up in correctional facilities was evidence that "there is a hereditary thread" and that any grandchildren born of these boys would certainly be defective.[60]

In order to express his opinion about Butler's request, Mr. Romero went to talk to the health officer at the San Francisco Detention Hospital who had admitted Juan to the Preston School of Industry. According to the health officer, Mr. Romero was "violently opposed" to Juan's sterilization and rejected the claim that he was feebleminded. In the end, racial discrimination against Mr. Romero and his boys converged with Butler's legal authority to sterilize without consent. As the health officer explained to Butler, Mr. Romero "is, of course, an ignorant, unintelligent Spanish man, and it is impossible to convince him of the value of the operation for sterilization either for his son's protection or for that of society. We hope that he will not give you a lot of trouble if you find it necessary to perform the operation at some future date without his consent in order to be able to parole" Juan.[61] Six months after this letter exchange, in March 1932, Butler convened a conference on Juan's case, and decided that the presence of three defectives in one family and the thirteen burglaries attributed to Juan warranted his sterilization, which shortly thereafter was approved in Sacramento.

In addition to highlighting parental interventions in sterilization decisions, Juan's case points to an important aspect of eugenic sterilization in California, namely, the disproportionate number of operations carried out on those with Spanish surnames, most likely persons of Mexican descent, in the state's feebleminded facilities. Analysis of data from Sonoma and Pacific Colony for the period from 1936 to 1949 reveals that Spanish surnamed patients represented an average of 18.5 percent and 26 percent, respectively, of patients sterilized. These figures are striking given that admission rates over this 13-year period ranged from below 1 percent to only as high as 3 percent at Sonoma and Pacific Colony, a discrepancy that clearly demonstrates that Spanish surnamed patients were targeted for reproductive surgery. Esperanza represented in multiple ways the patients sterilized at Pacific Colony. She was Mexican American, classified as an imbecile, and described as a "destructive girl who has become a social problem."[62]

Juan's story also captures the fervor of Butler, who ran Sonoma from 1918 to 1949 and proudly performed approximately 4,000 surgeries. Just two years after assuming the superintendence of Sonoma, Butler, endorsing the 1917 sterilization law (particularly its stipulation that sterilization be a precondition for discharge), declared that "all defectives who are capable of propagating, especially the hereditary class," should "be asexualized before leaving the institution."[63] Butler also was convinced that combating this menace required extending California's sterilization program beyond state institutions. By turning Sonoma into something of a revolving operating room, he made great strides toward this goal. Butler did what he could to ensure that teenage girls identified as unruly, promiscuous, and mentally defective by case workers and county officials were temporarily sent to Sonoma for salpingectomy. This pattern was documented in 1926 by the prominent California eugenicist Paul Popenoe during a site study of Sonoma conducted for the Pasadena-based Human Betterment Foundation. Popenoe noted, "It appears that something like 25% of the girls who have been sterilized were sent up here solely" for surgery. "They are kept only a few months—long enough to operate and install a little discipline in them; and then returned home."[64] According to Butler, "sterilization only" cases comprised 21 percent of Sonoma's load, and it was routine for persons categorized as "retardates" (possessing an IQ of 80 or below) to be surgically fixed and released in under a month's time.[65] Juan, who was transferred from Preston for the purposes of sterilization, demonstrates that temporary confinement for "sterilization only" also impacted boys.

Superintendents like Butler at Sonoma, Webster at Patton, and Smyth at Stockton, all of whom occupied their posts from the 1920s to the 1940s, helped to put the Golden State at the vanguard of eugenic sterilization. The counterexample of Dr. Leonard Stocking, who ran California's Agnews insane hospital from 1903 to 1931, reinforces this point. Stocking was much more cautious than many of his colleagues with regard to sterilization. For example, in response to the 1916 questionnaire sponsored by the California Board of Charities and Corrections, Stocking replied that he had performed practically no sterilizations, principally because he did not "think direct benefit to the patient is to be expected unless it may be in cases where the mental trouble follows and recurs with pregnancy or childbirth."[66] His reticence is evident in the official statistics, which show comparatively fewer operations performed at Agnews during Stocking's superintendence and their twofold increase after his retirement.[67]

Indiana and California ushered in America's sterilization era in the early 1900s. In distinct ways each state played a decisive role in the history of eugenic sterilization. Indiana passed the first law in the country, indeed in the world, although it soon became defunct, eventually was ruled unconstitutional, and two decades after passage was replaced with a more subtle act that nonetheless resembled sterilization laws in other states. While California could not claim the title of first, it unquestionably held the distinction of most, performing more than twice as many sterilizations as its closest rival, Virginia (approximately 8,000). One of the primary reasons for the Golden State's tremendous number of reproductive surgeries was that, unlike Indiana, its law was never seriously contested in the courts. The permanence of California's act permitted superintendents, who operated with little scrutiny, to make reproductive surgery central to the modus operandi of their institutions.

Within two decades, divergence had supplanted the initial convergence of sterilization policies in Indiana and California. Yet by the 1950s the two states resumed parallel tracks. Surgeries decreased substantially in the early to mid-1950s due, in California, to a dramatic revision of the law, and in both states, to the administrative reorganization of mental health agencies and a far-reaching shift in attitudes toward retardation and mental illness. As with other aspects of American eugenics, such as antimiscegenation and racially guided immigration laws, sterilization statutes were not purged from the books until the 1960s and 1970s. For example, Indiana's legislature repealed the 1927 law in 1974, and California followed suit five years later, expunging the 1909 act. In each instance, many legislators were stunned to learn that sterilizations were still performed sporadically, and they were anxious to get rid of laws that seemed legally crude and socially antiquated.

Ultimately, sterilization statutes in California and Indiana were translated into surgeries that terminated the reproductive capacity of patients and inmates committed to a variety of state institutions. Juxtaposing the lived experiences of sterilization in both states reveals significant differences in terms of affected populations. Whereas in California, more operations were performed on the mentally ill than the feebleminded, in Indiana the feebleminded were sterilized to the near exclusion of the epileptic and insane, even though the law was designed to apply to all these groups. Beyond institution or diagnosis, what remained constant was the vulnerability of those sterilized, many of whom were minors whose parents or adult guardians either welcomed or acquiesced to the medical paternalism of superintendents. Again and again, sterilization cases from both places reveal that once youths became ensnared in the juvenile system, their lives—the future of their procreative capacity—often lay in the hands of experts with biased and categorical ideas about who was defective and diseased.

Bringing patient stories to the forefront underscores that rather than a simple story of victimization, eugenic sterilization involved complex human actors whose lives dramatically intersected with some of the more egregious chapters in the modern history of mental health, disability, reproduction, and the state. However, without access to a deeper evidence base, our knowledge of these largely forgotten and silenced historical actors is likely to remain quite restricted, not to mention highly filtered through the bureaucratic and administrative scripts of institutional documents. Even with these limitations, we can glean some semblance of the trials and tribulations of those sterilized in California and Indiana, and situate their experiences along the wide-ranging continuum of the eugenic practices and policies that held forth in the United States for much of the twentieth century.

NOTES

1. "Indiana Sterilization Hearing, 1950, in the Case of ____, Medical Department, Fort Wayne State School, Fort Wayne, Indiana," Fort Wayne State School, Patient Files, Indiana State Archives (ISA). Names have been changed to protect the privacy of patients.

2. "Recommendation and Approval for Vasectomy or Salpingectomy for the Purpose of Sterilization," Pacific Colony, 1950. This is one of approximately 15,000 sterilization orders contained in microfilm reels housed at the California Department of Mental Health, accessed per IRB approval for Project #07-06-51, from the Committee for the Protection of Human Subjects, California Health and Human Services Agency. Names have been changed to protect the privacy of patients.

3. See sterilization numbers and statistics in Lisa M. Matocq, ed., *California's Compulsory Sterilization Policies, 1909–1979, July 16, 2003, Informational Hearing,* California legislative report of the Senate Select Committee on Genetics, Genetic Technologies, and Public Policy (December 2003).

4. See, for example, James W. Trent Jr., *Inventing the Feeble Mind: A History of Mental Retardation in the United States* (Berkeley: University of California Press, 1994); Steven Noll and James W. Trent Jr., eds., *Mental Retardation in America: A Historical Anthology* (New York: New York University Press, 2004), 281–99; Joel Braslow, *Mental Ills and Bodily Cures: Psychiatric Treatment in the First Half of the Twentieth Century* (Berkeley: University of California Press, 1997); Wendy Kline, *Building a Better Race: Gender, Sexuality, and Eugenics from the Turn of the Century to the Baby Boom* (Berkeley: University of California Press, 2001).

5. See Harry H. Laughlin, *Eugenical Sterilization in the United States* (Chicago: Psychopathic Laboratory of the Municipal Court of Chicago, 1922); Philip R. Reilly, *The Surgical Solution: A History of Involuntary Sterilization in the United States* (Baltimore: Johns Hopkins University Press, 1991).

6. On Sharp's initial rationale for vasectomy, see Elof Axel Carlson, *The Unfit: A History of a Bad Idea* (Cold Spring Harbor, N.Y.: Cold Spring Harbor Laboratory Press, 2001), and Angela Gugliotta, "'Dr. Sharp with His Little Knife': Therapeutic and Punitive Origins of Eugenic Vasectomy—Indiana, 1892–1911," *Journal of the History of Medicine* 53:4 (1998): 371–406.

7. Harry C. Sharp, *Vasectomy: A Means of Preventing Defective Procreation* (Jeffersonville: Indiana Reformatory Print, 1909), 6.

8. "An Act Entitled to Prevent Procreation of Confirmed Criminals, Idiots, Imbeciles, and Rapists," *Laws of the State of Indiana* (Indianapolis: Wm. B. Buford, 1907), 377–78. For a political and legislative analysis of Indiana's sterilization laws, see Jason Lanzter, this volume.

9. "Surgeons to Deal with Criminals," *Indianapolis Morning Star*, March 7, 1907, 10.

10. Sharp, *Vasectomy*, 2–3.

11. See Gugliotta, "'Dr. Sharp with His Little Knife.'" Figures for 1908 were calculated by archivist Vicki Casteel, Sterilization File, ISA.

12. See Correspondence Folder, Thomas R. Marshall Papers, ISA; Gugliotta, "'Dr. Sharp with His Little Knife'"; Carlson, *The Unfit*, 207–22.

13. See Reilly, *The Surgical Solution*; Laughlin, *Eugenical Sterilization*.

14. George A. H. Shideler to Governor J. P. Goodrich, September 9, 1909, folder 2, box 159, Indiana Reformatory Correspondence, Documents and Reports, Papers of Governor James Goodrich, ISA.

15. *Williams v. Smith*, No. 23,709, State of Indiana in the Supreme Court, Brief of Appellee, 4–5; *Smith v. Williams*, No. 12,106, Appeal from the Clark Circuit Court to the Supreme Court of Indiana, Sterilization File, ISA; "Eugenic Sterilization in Indiana," *Indiana Law Journal* 38:2 (1963): 275–89.

16. See *Sixty-first Annual Report of the Fort Wayne State School, Fort Wayne, Indiana, for the Fiscal Year Ending June 30, 1939* (Indianapolis: Wm. B. Burford Printing Co., 1939), 5. Nearly every annual report included a brief historical profile of the institution.

17. See *Fifty-third Annual Report of the Fort Wayne State School (also Twelfth Annual Report of the Muscatatuck Colony) for the Fiscal Year Ending September 30, 1931* (Indianapolis: Wm. B. Burford Printing Co., 1932).

18. Ibid., 12; *Fifty-fourth Annual Report of the Fort Wayne State School (also Thirteenth Annual Report of the Muscatatuck Colony) for the Fiscal Year Ending September 30, 1931* (Indianapolis: Wm. B. Burford Printing Co., 1933), 35.

19. Laughlin, *Eugenical Sterilization*, 17; on Hatch, see Braslow, *Mental Ills and Bodily Cures*, 56.

20. Laughlin, *Eugenical Sterilization*, 18, 19.

21. *Third Biennial Report of the Department of Institutions of the State of California, Two Years Ending June 30, 1926* (Sacramento: California State Printing Office, 1926), 92.

22. Braslow, *Mental Ills and Bodily Cures*, 56.

23. See Stern, *Eugenic Nation: Faults and Frontiers of Better Breeding in Modern America* (Berkeley: University of California Press, 2005).

24. See Paul A. Lombardo, *Three Generations, No Imbeciles: Eugenics, the Supreme Court, and Buck v. Bell* (Baltimore: Johns Hopkins University Press, 2008).

25. Reilly, *The Surgical Solution*, 97–101.

26. See *Statistical Report of the Department of Institutions of the State of California, Year Ending June 30, 1935* (Sacramento: California State Printing Office, 1936); *Statistical Report of the Department of Institutions of the State of California, Year Ending June 30, 1942* (Sacramento: California State Printing Office, 1943), 98; Matocq, ed., *California's Compulsory Sterilization Policies.*

27. See, for example, *Seventh Biennial Report of the State Commission in Lunacy for the Two Years Ending June 30, 1910* (Sacramento: Superintendent State Printing, 1910), 19–23; *Third Biennial Report,* 93–94; Fred O. Butler, "A Quarter of a Century's Experience in Sterilization of Mental Defectives in California," reprint from the *American Journal of Mental Deficiency* (1945), contained in Matocq, ed., *California's Compulsory Sterilization Policies.*

28. Although cases involving allegations of medical malpractice, demands for damages, and petitions for the state to provide nontherapeutic surgical sterilizations for the indigent were heard in California's supreme and appellate courts between 1930 and 1979, only *Sara Rosas Garcia v. State Department of Institutions* challenged the constitutionality of the sterilization law. See *Jessin v. County of Shasta,* Civ. No. 12,027, Court of Appeal of California, Third Appellate District, 274 Cal. App. 2d 737; 79 Cal. Rptr. 359; 1969 Cal. App. LEXIS 2107; 35 A.L.R.3d 1433, July 11, 1969; Kline, *Building a Better Race,* chap. 4.

29. Petition for Writ of Prohibition of Mandate, Civ. 12533, in the District Court of Appeal of the State of California, Second Appellate District, *Sara Rosas Garcia vs. State Department of Institutions of the State of California et al.,* December 13, 1939, on file at the California State Archives. An abridged version is available through Lexis-Nexis: see *Sara Rosas Garcia v. State Department of Institutions,* Civ. No. 12,533, Court of Appeal of California, Second Appellate District, Division One, 36 Cal. App. 2nd 152; 97 P.2d 264; 1939 Cal. App. LEXIS.

30. See Legislative History, Senate Bill 750, microfilm 3:2 (4); "Legislative Memorandum," April 4, 1953, Legislative History, Assembly Bill 2683, microfilm reel 3:2 (10); Frank F. Tallman to Honorable Earl Warren, March 31, 1953, Legislative History, Assembly Bill 2683, microfilm reel 3:2 (10), CSA.

31. For a discussion of Catholic opposition to sterilization, which solidified in the 1920s, see Christine Rosen, *Preaching Eugenics: Religious Leaders and the American Eugenics Movement* (New York: Oxford University Press, 2004).

32. On Earl Warren's gubernatorial administration in California and his rise to Supreme Court justice, see Jim Newton, *Justice for All: Earl Warren and the Nation He Made* (New York: Riverhead Books, 2006).

33. See "Background Paper," in Matocq, ed., *California's Compulsory Sterilization Policies.*

34. "Background Paper," and "Sterilization Operations in California State Hospitals, April 26, 1909, through June 30, 1960," in Matocq, ed., *California's Compulsory Sterilization Policies.* Also see "History, Description, and Evaluation of the Department of Mental Hygiene," (1962), Papers of Nathan Sloate, Department of Mental Hygiene Records (DMH), CSA.

35. "Sonoma State Hospital," Biennial Report, 1952–54, Sonoma State Hospital, Administrative Files, DMH, F3501, Loc D0419, box 154, CSA.

36. Not until the 1940s, when California claimed about 60 percent of all operations nationwide, did a few places, such as Delaware, North Carolina, and Virginia, begin to consistently overtake California in per capita or annual terms. Figures derived from

"U.S. Maps Showing the States Having Sterilization Laws in 1910, 1920, 1930, 1940," *Publication No. 5* (Princeton: Birthright, n.d.), in Matocq, *California's Compulsory Sterilization Policies;* Clarence J. Gamble, "Preventive Sterilization in 1948," *JAMA* 141:11 (1949): 773; Gamble, "Sterilization of the Mentally Deficient under State Laws," *American Journal of Mental Deficiency* 51:2 (1946): 164–69. Delaware was the only state that outpaced California in per capita terms in the 1930s, with a rate ranging between about 80 and 100 sterilizations per 100,000 individuals.

37. "Eugenic Sterilization in Indiana" notes inconsistent interpretations of criminal immunity; "An Act to Provide for the Sexual Sterilization of Inmates" (1927), Sterilization File, ISA.

38. See Fort Wayne Minutes, 1931–47, ISA.

39. L. Potter Harshman, "Medical and Legal Aspects of Sterilization in Indiana," *Journal of Psycho-Asthenics* 39 (June 1933–June 1934): 189.

40. See "Eugenical Sterilization in Indiana."

41. Sonoma State Hospital, interoffice memorandum, November 29, 1956, re: W & I Code Changes, Sonoma State Hospital, Administrative Files, DMH, F3501, Loc D0419, box 153, CSA.

42. See Trent, *Inventing the Feeble Mind,* and Noll and Trent, *Mental Retardation in America;* for an elaboration of this attitudinal change in California, see "Statement of Progress, Department of Mental Hygiene for the Period 1943–1950," November 1, 1950, DMH, Papers of Nathan Sloate, CSA.

43. "Sterilization Operations Performed, Patton State Hospital, during Month of March 1939." Her operation occurred on March 23, 1939.

44. "Recommendation and Approval for Vasectomy or Salpingectomy for the Purpose of Sterilization," Patton State Hospital, 1939.

45. *Statistical Report of the Department of Institutions of the State of California, Year Ending June 30, 1939* (Sacramento: California State Printing Office, 1939), 26.

46. John A. Reily to John Randolph Haynes, February 19, 1916, box 193, Haynes Papers, 1241, Special Collections, UCLA.

47. *Third Biennial Report of the State Department of Institutions* (1926), 83.

48. To be true to the historical era of eugenic sterilization, I use the terms employed at the time. Today these kinds of facilities generally are referred to as developmental centers (feebleminded homes) and state hospitals (mental hospitals).

49. *Statistical Report of the Department of Institutions of the State of California, Year Ending June 30, 1940* (Sacramento: California State Printing Office, 1940), 12.

50. Braslow, *Mental Ills and Bodily Cures,* 21, 80.

51. *Biennial Report* (1939), 26.

52. On Smyth's ardent support of sterilization, see Margaret L. Smyth to John Randolph Haynes, June 3, 1916, box 17, Haynes Papers, 1241, Special Collections, UCLA; Smyth, "Psychiatric History and Development in California," *American Journal of Psychiatry* 94 (1938): 1223–36.

53. See Stern, *Eugenic Nation,* chap. 3.

54. Sterilization order in the form of a letter, Butler to Earl E. Jensen, March 22, 1928.

55. Sterilization order in the form of a letter, Butler to Dr. J. M. Toner, January 9, 1932.

56. "Indiana Sterilization Hearing, 1950, in the Case of ____, Muscatatuck State School, Butlerville, Indiana," April 16, 1942, Muscatatuck State School, Patient Files, ISA.

57. See entries for 1942 in Fort Wayne State School Minutes, vol. 1, 1931–47, ISA.

58. See entries for 1953 in Fort Wayne State School Minutes, vol. 2, 1948–55, ISA.

59. For insight into California's youth of color in state institutions, see Miroslava Chávez-García, "Intelligence Testing at Whittier School, 1890-1920," *Pacific Historical Review* 76:2 (2007): 193–228.

60. Butler to _____, November 21, 1931, Inmate #13694, F3738:20, Inmate Records, Inmates Histories, Papers of the Preston School of Industry, California Youth Authority, CSA, consulted per confidentiality agreement signed at the CSA.

61. J. C. Geiger to Butler, November 24, 1931, Inmate #13694, ibid.

62. "Recommendation and Approval for Vasectomy or Salpingectomy for the Purpose of Sterilization," Pacific Colony, _____, 1950.

63. Fred O. Butler, "Report of Medical Superintendent of the Sonoma State Home," *First Biennial Report of the Department of Institutions,* 1920, 80.

64. Paul Popenoe to Ezra S. Gosney, March 25, 1926, box 7, folder 2, Papers of Ezra S. Gosney and the Human Betterment Foundation, Institute Archives, California Institute of Technology.

65. See interview of Fred O. Butler; Butler, "A Quarter of a Century's Experience."

66. Leonard Stocking to John Randolph Haynes, March 11, 1916, box 193, Haynes Papers, 1241, Special Collections, UCLA.

67. See "California's Compulsory Sterilization Policies"; Braslow, *Mental Ills and Bodily Cures.*

SIX

Eugenics and Social Welfare
in New Deal Minnesota

MOLLY LADD-TAYLOR

Tena and Stewart had "many good traits and were fond of their children," but social workers found their living conditions "impossible." It was 1937, and the couple "could not cope" with the "existing conditions" of depression and unemployment. He drank, and she was "stepping out"—despite being pregnant with her fourth child. Their children appeared neglected and abused. After IQ tests found both parents to be feebleminded,[1] they were committed to state guardianship, sent to the state institution, and sterilized. Stewart's parents quickly consented to his operation, and he returned home after two months, but Tena's parents "caused" her operation to be delayed. She was reunited with her family eight months later, but the couple remained wards of the state. A social worker helped Stewart get New Deal work relief and kept in "close touch" with Tena as she reestablished her home.[2]

Tena and Stewart were among the approximately 1,200 Minnesotans sterilized during the Great Depression.[3] Their welfare dependency and problematic parenting brought them to the attention of the county child welfare board, and their low IQ scores provided proof of their mental deficiency. Their case, which was unusual only because both husband and wife were sterilized, was presented to the American Association for Mental Deficiency as an example of successful casework. The write-up made no reference to heredity or eugenics, however; the point was that the couple's economic circumstances and parenting skills could be improved with positive social work intervention and surgery. Eugenic sterilization, in this case, was a routine decision aimed at ensuring that a welfare-dependent family did not have any more mouths to feed and that their children did not get into trouble. Their surgery was propelled less by a eugenics-based "quest for racial purity" than by specific local concerns about welfare dependency and social disorder.[4]

This essay examines the routine operation of eugenic sterilization as a social welfare policy during the New Deal, the period of its greatest activity. Although Minnesota's sterilization law was enacted in 1925 and stayed on the books for 50 years, nearly 45 percent of the state's recorded sterilizations were performed between 1933 and 1940, mostly on white working-class women considered to be feebleminded. Sterilizations dropped off during World War II because of a shortage of medical and nursing personnel, and the program never returned to its earlier scale. By 1963, the last year for which statistics are available, at least 2,350 Minnesotans—78 percent of them women—had been sterilized, but Minnesota accounted for just 4 percent of all sterilizations in the nation.[5]

Many historians have commented on the relationship between eugenics and European welfare states, but little work has been done on the relationship between eugenics, liberal social welfare policy, and day-to-day sterilization routines in the United States.[6] Minnesota is a good site for investigation into this topic because its eugenic sterilization law was considered voluntary (consent of kin was required), and it was administered within a child welfare system that won national praise. Like public welfare in general, eugenic sterilization was shaped by the social-control goals of reducing dependency and disorder, and characterized by local variation, political manipulation, and a gap between rhetoric and routinization. Eugenics-inspired fears of the menace of the feebleminded were clearly central to the passage of the 1925 sterilization law, but the bill's actual administration over the years had as much to do with fiscal and welfare politics as with an attempt to improve the human race. The eugenic and welfare functions of Minnesota's sterilization program overlapped when the object of eliminating the unfit converged with keeping taxes and relief costs low, yet at times the welfare goal of providing assistance to the poor and the eugenics goal of reducing their numbers collided.[7]

The legal foundation for Minnesota's eugenic sterilization law was laid in its Children's Code, a package of 35 laws passed in 1917 to modernize the state's child welfare system. According to one social work historian, the code launched a "new era of child welfare work" and made Minnesota "one of the leading states measured by its children's laws."

It modernized adoption procedures, extended the rights of illegitimate children, revised the state's mothers' pensions and juvenile court laws, and created an administrative apparatus, consisting of a state children's bureau and county child welfare boards, to take children's services into every corner of the state. Most important for the history of sterilization, the Children's Code included a civil commitment law that empowered county probate judges to commit neglected, dependent, and delinquent children—and any person "alleged to be Feeble Minded, Inebriate, or Insane," regardless of age—to state guardianship without the approval of parent or kin. The guardianship was for life, unless the person was specifically discharged. Once committed as feebleminded, a ward took on the status of a permanent child: he or she was unable to vote, own property, manage his or her financial affairs, or marry without the state's approval. The Board of Control, as legal guardian, decided whether the ward should be placed in an institution or stay in the county under the supervision of the child welfare board.[8]

The fact that a compulsory commitment law for so-called defectives was part of the Children's Code reveals the deep intellectual and administrative connections between eugenics and child welfare in Progressive Era Minnesota. It also demonstrates the enormous influence of Dr. Arthur C. Rogers, the esteemed superintendent of the Minnesota School for the Feebleminded from 1885 to 1917. Although Rogers died before either compulsory commitment or eugenic sterilization became law, his three-pronged strategy for the identification and control of the feebleminded—eugenic family studies, mass IQ testing, and compulsory civil commitment—shaped Minnesota's eugenics policy into the 1930s. In 1910, the year the Eugenic Record Office (ERO) was founded and just two years after Alfred Binet developed the first modern intelligence test, Rogers arranged for two ERO field workers to study the family histories of students at the Faribault School. He also hired psychologist Frederick Kuhlmann, a former classmate of Lewis Terman at Clark University, to conduct IQ tests at the institution (and throughout the state). Rogers's advocacy of civil commitment as a eugenics strategy was especially innovative, for it permitted authorities to institutionalize supposedly defective women and girls without their families' consent. It is a testament to Rogers's stature—and to the general influence of his eugenics philosophy on Minnesota reformers—that the Minnesota Child Welfare

Commission wrote a compulsory commitment law for defectives into the Children's Code.[9]

Rogers's welfarist approach to eugenics also influenced Minnesota sterilization policy. Although the Faribault superintendent opposed a 1913 bill that would have sterilized criminals as well as the feebleminded, he registered general approval of the sterilization idea, and the 1925 sterilization law was consistent with his careful approach.[10] Unlike the better-known laws of Virginia, Indiana, and California, Minnesota's sterilization program applied only to persons who had been committed to state guardianship as feebleminded or insane. The sterilization of a feeble-minded ward could be authorized after "careful investigation of all the circumstances of the case," consultation with three experts (the superintendent of the School for the Feebleminded, a "reputable" physician, and a psychologist), and the written consent of the spouse or nearest kin. If no relative could be located, the State Board of Control as legal guardian could give consent. A mentally ill person committed to the custody of the superintendent of the state hospital for the insane could be sterilized only if he or she had been a patient in the institution for six consecutive months. Both the patient and the next of kin had to give their written consent. It is not surprising, given the statutory requirements pertaining to institutionalization and personal consent, that fewer than 20 percent of sterilizations in Minnesota were performed on the insane.[11]

The administrative structure of Minnesota's sterilization program was also singularly welfare oriented. While most states placed the authority to order sterilizations in a state eugenics board or the board of trustees or superintendent of a state institution, Minnesota made the State Board of Control its chief sterilization agency. A three-member panel appointed by the governor to "secure the economical management" of state institutions, the Board of Control oversaw the operation of mental hospitals, prisons, poorhouses, the state orphanage, and the school for the feebleminded; it also assisted county child welfare boards with the care of dependent or neglected children and distributed emergency relief during the New Deal.[12]

If the administrative structure of the sterilization program reflected its overlapping welfare and eugenic functions, the inherent tensions between these two aims was evident in the simmering dispute between state welfare officials and the Minnesota Eugenics Society. The founding meeting of the eugenics society was attended by several prominent welfare reformers, including Minneapolis settlement worker Catheryne Cooke Gilman, the executive secretary of the Women's Cooperative Alliance and a member of the Child Welfare Commission that drafted the Children's Code. Yet there is little evidence of sustained interaction. No member of the Child Welfare Commission sat on the council of the Minnesota Eugenics Society when it was founded in 1923. A few high-profile men, including Frederick Kuhlmann, George G. Eitel (who would eventually perform many of the surgeries at Faribault), and Minneapolis General Hospital superintendent Walter List joined the Minnesota Eugenics Society, but it was Charles Dight, the organization's eccentric president, who became the public face of eugenics in Minnesota.

Dight, a socialist physician and former Minneapolis alderman who lived for a while in a tree house, was a tireless eugenics campaigner. He believed fervently that the "socially unfit"—the insane, the diseased, the feebleminded and epileptic, the criminally inclined, and the avaricious— were a "peril to the nation," and he was dogged in his pursuit of a eugenics law. Dight's efforts on behalf of a sterilization bill were publicly acknowledged with an invitation to attend the first six operations on January 8, 1926.[13]

At first glance, Dight's high-profile career supports the popular view of eugenic sterilization as the product of a racist campaign by scientific and medical elites to "create a superior Nordic race."[14] Dight worried about the influx and "rapid reproduction" of the "mentally inferior of Europe," complained that Minnesota's "restricted" sterilization law did not cover Mexican migrant workers, and wrote an admiring letter to Hitler in 1933. Determined to extend "eugenic" sterilization to rapists, three-time felons, undesirable immigrants, and others who were eugenically unfit but not committed to state guardianship, Dight waged a loud but ultimately unsuccessful campaign to make Minnesota's sterilization law more like California's. He was vociferous in his criticism of the casework orientation of the Board of Control (especially its sole female member, Blanche La Du) and lobbied stubbornly for an office of state eugenist or county sterilization boards to replace it as the state's chief sterilization authority.[15]

Yet the Board of Control consistently rejected Dight's sweeping eugenics vision in favor of a social welfare policy of "selective sterilization" in individual cases. La Du reportedly called Dight's far-reaching proposals dangerous and hung up the telephone or walked away when he tried to argue with her. Another Board of Control employee recalled ducking out a side door so she would be "not in" whenever he visited her office.[16] Both sides agreed that mental defect was largely inherited and posed a burden on the welfare system, yet while Dight accentuated the eugenic benefits of sterilization, the Board of Control emphasized the need for individual casework and public welfare. To state welfare officials, the most compelling justification for sterilization was not eugenic, but "socio-economic . . . that the feeble-minded parent cannot provide a stable and secure family life for his children."[17]

Scientific and medical elites played an important role in lobbying for a sterilization law and performing the actual surgeries, but they played a surprisingly small role in the day-to-day decisions about who would be sterilized in Minnesota. Under the 1917 commitment law, any family member or "reputable" citizen living in the same county as an alleged defective could initiate a commitment proceeding, and a local probate judge had the sole authority to decide if the petition had merit and should continue. The State Board of Control was permitted to send a "person skilled in mental diagnosis" to examine the alleged defective and attend a hearing, but the expert played only a consultative role. The actual finding of feeble-mindedness was made by a board of examiners comprised of the probate judge and two licensed physicians appointed by the judge—although if a person were "obviously feebleminded," the judge could dispense with the board of examiners and make the decision on his own. Thus the first and most significant step in the sterilization process—the power to diagnose and commit a feebleminded individual to state guardianship—rested in the hands of a probate judge, an elected official not required to have any legal or medical training. (If a feebleminded ward petitioned to be "restored to capacity," the case would be heard by a district court judge, presiding without a jury.)[18]

Unquestionably, the decision to commit was the crucial moment in the administration of the sterilization law. Probate judges had a great deal of latitude in making this decision because the statutory definition of a "feebleminded person"—as someone "who is so mentally defective as to be incapable of managing himself and his affairs, and to require supervision, control and care for his own or the public welfare"—was so imprecise.[19] Most judges relied on the standard markers of feeblemindedness—visual signifiers, such as an evident physical disability or a "large tongue" or "protruding lip"; social characteristics, such as poor English language skills, a weak school or employment record, or a disorderly home; or behavioral indicators such as alcoholism or "sex delinquency."[20] Many judges also considered the results of intelligence tests, although these were often given under "conditions that were far from ideal—in a crowded room, in a home with a child often pounding on the door, in the yard or in the car"—a fact that surely added to the numbers of working-class Minnesotans determined to be feebleminded. Many of the personal traits taken as evidence of hereditary feeblemindedness in the 1930s were seen by later generations as the effects of a culture of poverty—fatalism, inability to delay gratification, and a low level of aspiration. The feebleminded "lack common sense, foresight, are unable to resist ordinary temptations, act on impulse, and have little or no initiative," the control board explained in a widely distributed memo. "They have about the same desires as normals, including sexual, but lack ability to control them. They usually have poor homes."[21]

The connection officials saw between feeblemindedness, poverty, and sexual misbehavior was not simply due to their frame of mind; it was rooted in the fiscal and administrative structures of Minnesota's welfare system. Probate courts doubled as juvenile courts in Minnesota's 84 non-urban counties, and rural judges routinely encountered cases of incorrigibility, truancy, sex delinquency, and child neglect—all highly subjective offenses that were associated with feeblemindedness. In addition, unpaid child welfare board members, who conducted mothers' aid investigations and arranged for foster-care or adoption placements for needy children, were inclined to stress "the relationship of low mentality to social problems." Poor relief was funded through local property taxes, and cash-strapped child welfare boards worried about the pauperizing effects—and rising costs—of public relief. They often treated destitute families as a burden on taxpayers and saw hereditary feeblemindedness as the cause of poverty, sexual misconduct, and crime. Many welfare workers embraced Minnesota's three-step program of committing, institutionalizing, and sterilizing the feebleminded because it shifted part of the economic and social responsibility for their most troublesome charges from the county to the state.[22]

While feebleminded commitments were made (and occasionally unmade) at the local level by hundreds of different judges, the decision to sterilize was centralized in St. Paul. The Board of Control gave final approval, but the central figure in Minnesota's sterilization program was Mildred Thomson, who ran the control board's bureau for the feeble-minded and epileptic from 1924 to 1959 and was de facto guardian of the state's feebleminded wards. An Atlanta native, Thomson had a master's degree in education from Stanford University, where she wrote a thesis on the IQ testing of schoolchildren with the famous psychologist Lewis Terman. Yet she saw herself primarily as a social worker. In her memoirs, she made a point of distinguishing the attitudes of social workers "actually working with individuals" from the men, such as Frederick Kuhlmann and Charles Dight, whose interest in eugenics she saw as more abstract and indiscriminate. Throughout her long career, Thomson stressed the state's "obligation" to the feebleminded, who "even though grown . . . are really children" in need of protection. Like her superiors at the Board of Control, she opposed the "overzealous conclusion that sterilization of the feebleminded is wholly justified by the eugenic factors," but supported "selective sterilization, with consideration of the individual tempering each selection" when it was in the "best interests" of the ward. "WE ARE PRIMARILY THE FRIENDS OF THE FEEBLE-MINDED," Thomson reminded county child welfare board members in a widely distributed memo, "MADE SO BY LAW AND ALSO BY OUR NATURAL SYMPATHIES, I BELIEVE."[23] After the war, Thomson's "natural sympathies" for the feebleminded and their families led her to play a major role in the 1950 founding of the national Association for Retarded Children, a parent advocacy group for people with intellectual disabilities; her contribution to ARC is still noted on the organization's website. The relative ease with which Thomson made the transition from eugenics administrator to reform facilitator reveals both the constancy of her welfare-casework orientation and the mutability of her social work role.[24]

Thomson's view of surgical sterilization as just one part of a broad social work program to assist the feebleminded also helps to explain another apparent paradox: why sterilizations increased in the 1930s, just as scientific and popular critiques of eugenics were beginning to command attention. As the decade opened, two events shifted Minnesota's political and cultural landscape away from a general support for "eugenics":

Pope Pius XI issued his 1930 encyclical on marriage, which strengthened Catholic opposition to eugenic sterilization, and the charismatic Swedish-American attorney Floyd Olson, a forceful advocate for emergency relief and social security, was elected the state's first Farmer-Labor governor. Amidst the turmoil of the Depression, impassioned battles over Farmer-Labor politics, unionization, and the New Deal swept away the erstwhile fixation on the unfit. The Minnesota Eugenics Society "faded out of existence," but routine sterilizations continued on.[25]

Most historians have attributed the nationwide increase in sterilizations during the Depression to a general acceptance of eugenics beliefs and budgetary pressures on the public purse: sterilizing and releasing the "high-grade" feebleminded freed up space in overcrowded state institutions.[26] Yet the increase in sterilizations in 1930s Minnesota coincided not with the "harsh economic realities" of the Depression but with the influx of federal funds during the New Deal. About 67 Minnesotans were sterilized annually between 1926 and 1929. The number of recorded sterilizations increased slightly in the early years of the Depression, but more than doubled—from 79 to 144—between 1932 and 1934. Nearly 140 Minnesotans were sterilized each year for the rest of the decade.[27]

The increase in sterilizations came partly because the New Deal funneled emergency relief through existing state agencies, vastly increasing the power of amateurish child welfare workers in the counties and the old-school social workers at the State Board of Control. In 1933, Governor Olson designated the Board of Control the State Emergency Relief Administration responsible for allocating emergency aid, distributing food and surplus commodities to drought-scarred counties, and administering the major New Deal work relief programs, the Civilian Conservation Corps and (after 1935) the Works Progress Administration (WPA). In 1937, when the state reorganized its welfare system to meet the administrative standards of the Social Security Act, the mostly volunteer child welfare boards established under the 1917 Children's Code were transformed into "county welfare boards" with higher budgets and greater investigative and economic powers. They retained their old responsibilities for child welfare and the state's feebleminded wards, but took on a colossal new job: the administration of "all forms of public assistance." The massive relief and social security programs dwarfed the control board's other programs. The program for the feebleminded "seemed especially overshadowed,"

Thomson recalled. "With few exceptions the welfare boards gave first consideration to other programs; supervision of the feebleminded was done almost solely on an emergency basis."[28]

The spectacular increase in federal welfare funding (and government jobs ripe for patronage) raised the stakes in the bitter partisan battles over taxation and relief. Traveling around the country for federal relief administrator Harry Hopkins, journalist Lorena Hickok reported that she had a "pretty poor impression" of relief work in Minnesota. "The whole set-up in this state was—pardon the vulgarity—lousy with politics." Although Olson was "about the smartest 'Red' in this country," the Democrats were weak, and "most of the money and the press are Republican and hostile." Charges of Communist influence in state government, and the internal struggles that divided Farmer-Laborites after Olson's sudden death from cancer in 1936, added to the turmoil. The years from 1936 to 1939 were a time of "change and also of considerable stress," Thomson recalled with characteristic understatement. There were deep divisions between "extremely conservative" social workers and "extremely radical" ones and a "sense of uneasiness" in the state agencies. The Republicans regained the governorship in 1938, and the new governor, Harold Stassen, embarked on a major reorganization of state government that brought the 1939 elimination of the Board of Control. Suspected Communists were purged from state government, and the control board was replaced by a Social Security Board with two new departments, Social Welfare and Public Institutions. Yet sterilizations did not stop.[29]

Despite the turmoil in the state bureaucracy, Minnesota's program for the feebleminded remained outside the political fray for much of this period. Thomson's bureaucratic disposition and detachment from party politics—and the fact that her bureau had no direct money grants—likely kept the program for the feebleminded from getting politicized. Yet the inattention to her program also signals ideas about disability, dependency, and gender that all three parties at every level of government shared. Historian Alice Kessler-Harris has remarked on what she calls a "gendered imagination" which positioned men as independent workers and women as dependent family members and constituted a crucial measure of "what was fair" at a time when the social welfare functions of the federal government were expanding. Despite being deeply divided over unions, taxation, and the New Deal, most Minnesotans held the same gendered ideas about male breadwinning and the dignity of work—ideas shaped by disability as well as by gender. Thus, while policy makers in Washington and St. Paul designed work relief, union rights, and unemployment insurance that preserved the dignity of "employable" but temporarily jobless (white) men, public assistance for "unemployables" remained on an inferior track. Social assistance for single mothers, dependent children, and "defectives" (who, according to Thomson, were rarely "employable" under the "present industrial set-up") remained largely means- and morals-tested and locally funded and controlled. Perhaps this is why Olson, despite expressing little interest in the state's program for the feebleminded, signed a bill long advocated by Kuhlmann that authorized a statewide census of the feebleminded in 1935. The bill was more symbolic than practical, however, for no appropriation was ever made.[30]

For reasons both cultural and structural, Minnesota's guardianship program for the feebleminded received bipartisan support in the 1930s, and Mildred Thomson could proudly claim that hers was the most "feebleminded conscious" state. Even during the Depression, many social workers saw a "definite relationship between the persons receiving direct relief and work relief from the state and federal governments, and feeble-mindedness." Some welfare boards ordered IQ tests for parents and children living "under deplorable conditions" and had entire families committed to state guardianship as feebleminded. "Apparently the boards were satisfied that they had taken some kind of action," Thomson later observed, for although counties still had to pay for relief or help wards find WPA jobs, the state assumed a portion of the economic responsibility. "Later some of these families were difficult to work with," she conceded; "not all those tested and committed to guardianship under the circumstances proved to be really feebleminded and their frustrating experiences made them resentful. Tests and decisions had been made too hurriedly." Commitments were made by the probate courts, however, and state officials were "powerless" to undo court decisions once they were made.[31]

Note that Thomson's observations concerned feebleminded commitments, which could be ordered without the consent of kin, but not sterilization, which required relatives to consent. Minnesota officials took sterilization consent seriously, and it was recorded, along with basic demographic information like birth date, IQ, and county of residence, in a medical record book of the first 1,000 sterilization operations performed at the Faribault School for the Feebleminded. It is interesting that both consent of kin, which was required by law, and personal consent, which was legally unnecessary, were recorded. Members of the immediate family—parents, spouse, or siblings—consented to 82 percent of sterilizations, and personal consent was obtained for more than 97 percent of surgeries.[32] Of course, consent could be coerced, as when parents put pressure on a daughter who was sexually active or had a baby out of wedlock, or when officials withheld release from the institution until consent to sterilization was secured. Thomson insisted that the state would never "make a bargain" in exchange for sterilization and that people who opposed sterilization because of "religious and moral convictions" would face "no pressure" to give consent, but she regularly advised child welfare workers to tell families that release from the state institution would proceed "more easily and satisfactorily" after sterilization. Surgery should not be considered a condition of discharge, she said, but an "aid to the reconstruction of their lives outside the institution."[33]

Despite its obvious limitations, Minnesota's statutory consent requirement provided crucial leverage for disadvantaged families. A significant number of families, especially Catholics, refused to consent to a sterilization operation, and given popular beliefs about the impact of vasectomy on male sexual performance, the consent provision might have been a factor in the relatively small percentage of sterilizations performed on men. Yet even when operations were eventually performed, the consent provision helped poor families negotiate with welfare officials. The case of Sam and Lena B provides a good illustration. The couple had two children when they were committed to state guardianship in 1934. Although the welfare board wanted the pair sterilized immediately, relatives objected, and three years passed before the couple was taken to Faribault for surgery. A daughter was born in the intervening period, and the couple's fourth child, a son, was born in the institution. The family was reunited a few months after the parents were sterilized, and the social worker reported that they were happy: "Sam said he thought he had the right size family and was glad that they did not have any more children," she explained. The couple resisted the state's sterilization plan when they had only two children, but they were content to stop at four.[34]

As the case of Sam and Lena suggests, some welfare-dependent families tolerated—and occasionally even sought—surgical sterilization as a form of birth control paid for by the state. A Twin Cities social worker, who considered sterilization "questionable" when it came to eugenics, admitted that it "has worked out well in families where there were already enough children and the mother and father were convinced that there should not be any more." The state archives contain many examples of women like Annie, the mother of ten children, who entered the institution "for sterilization" and returned home as soon as they recovered from surgery. In their cases, the eugenics aim of preventing the births of potentially feebleminded children, the welfare objective of providing (minimal) assistance to relief-dependent families, and poor women's desire for reproductive control converged. Yet state-funded sterilizations could only be performed on feebleminded or insane persons under state guardianship, so poor women seeking state-funded sterilization were left exceedingly vulnerable. Countless women got stuck in the institution when outbreaks of influenza or the measles caused surgical delays or when a family member would not sign the consent form. (The patient, of course, could not legally sign for herself.) By the end of the decade, Minnesota's commitment-institutionalization-sterilization program was winning few praises outside the Board of Control. But the operations continued.[35]

By the end of the 1930s, political turmoil, partisan charges of patronage and corruption, and constant criticisms of state government created a space for a more open rebellion. In the early years of the sterilization program, resistance was largely invisible; most people rebelled by refusing to sign sterilization consent forms or simply by running away. Few Minnesotans pursued the initial right to a hearing, and fewer still sued for "restoration to capacity." As the program expanded, however, the opposition grew increasingly vocal. Socialist feminist author Meridel Le Sueur wrote about single pregnant working-class women struggling to escape police matrons and social workers wanting to sterilize them. Except for Le Sueur and a handful of Farmer-Labor radicals, however, there was little evidence of concern about women's reproductive or human rights.[36]

Indeed, most public criticism of Minnesota's program for the feebleminded was motivated by political opposition to the leftist Farmer-Labor regime. A particularly damning report, which focused partly on misspending and administrative problems at the School for the Feebleminded, was released shortly after the Republicans regained control of state government in 1939. A joint committee investigating the "acts and activities of the various governmental departments," chaired by conservative legislator A. O. Sletvold, accused Superintendent Edward Engberg of using tax dollars to renovate his home and condemned the scandalous treatment of patients. Inmates did not have enough milk or butter, there were not enough guards for violent male inmates, trench mouth was epidemic, and broken bones were left untreated for days. The committee recounted the sad story of Blanche Harkner as an "outstanding example" of medical neglect, although we know from other sources that Harkner's injuries resulted from a desperate attempt to escape a scheduled sterilization operation in February 1939. Harkner climbed out a window at the hospital where sterilizations were performed and fell two and a half stories to the ground, shattering both legs and injuring her spine. More than six days passed before she was transferred to a Minneapolis hospital where she could receive proper medical care. Yet while the Sletvold committee expressed outrage over the delay in treating Harkner's fractured bones, it was silent on the question of her sterilization. Faribault officials reported 136 sterilizations in 1939, Republican Harold Stassen's first year as governor, but Sletvold maintained that the "evils" wrought by the Farmer-Laborites were already being remedied by the new administration. Clearly, sterilization was a bipartisan affair.[37]

While the Sletvold committee and other legislative investigations of the 1930s got a lot of press, they had little real impact; more serious allegations of incompetence and mismanagement were leveled by national authorities. Between 1939 and 1941 two major studies—a U.S. Public Health Service survey of mental hospitals and a review of Minnesota's mental health program by the American Public Welfare Association— raised "very serious questions" about the program for the feebleminded and especially Mildred Thomson. Both reports called attention to the inordinate power of untrained county welfare boards and questioned the wisdom of placing the administration of Minnesota's mental health program under the agency responsible for prisons and welfare.[38]

Thomson's beloved program of legal guardianship for the feebleminded came in for particularly harsh criticism. "Certain people in Minnesota have been very proud of the attempts that they have made to address the problem of feeblemindedness," Milton Kirkpatrick of the Public Welfare Association wrote derisively, but the entire program was flawed.

The state placed too much emphasis on the social problems caused by the feebleminded and failed to recognize that "higher-grade" mentally deficient individuals could be rehabilitated and made self-supporting. About 500 new commitments were made each year, but guardianship meant little beyond facilitating admission or discharge from the state institutions; it did nothing for feebleminded wards living in the community. There was "little justification for the state assuming 'en masse' responsibility for so many people ... when [the state] has so little to offer," he concluded.[39]

Kirkpatrick also condemned the "startling procedures" at the state institutions. Crucial decisions about admission and discharge were not made by physicians or the institutions' medical superintendents, but by the supervisor of the feebleminded (Thomson) and shaped largely by local welfare needs. Discharge depended on the willingness of the county of residence to provide economic and social support for the ward, yet because counties paid only $40 per year to support wards who were institutionalized, Kirkpatrick surmised that there was "little likelihood that they will stir themselves very violently" to help feebleminded persons return to the community, where the cost of supervision was much greater. Curiously, he did not mention sterilization.[40]

The Public Health Service, however, wrote disapprovingly that the "sterilization of defectives is a vigorously advocated procedure." It noted that some Minnesota officials hung on to the "crudely unscientific notion" that surgery could eliminate hereditary defect, but the more common argument was economic: "Sterilization for even one feeble-minded person may save the community the cost of supporting another one ... [and] supervision is much cheaper if the morals of the person supervised are not a matter of concern." Like Kirkpatrick, the Public Health Service expressed concern that state wards who returned to the community were not supervised by professionals trained to work with the mentally retarded, but by county welfare boards "utterly inadequate" to the task.[41]

Thomson was stunned by the criticism. The release of the Kirkpatrick report in January 1941 "gave me a shock!" she recalled in her memoirs; "It was more than critical—it was devastating—in its estimate of the program for the mentally deficient and epileptic." But she dismissed the criticism as a professional dispute between psychiatrists and psychiatric social workers who wanted to "re-create the feebleminded in the image of the mentally ill" and those (like herself) who were more realistic. Kirkpatrick had little understanding of the extent and permanence of the feebleminded problem, Thomson complained; "he was assuming that if you refused to recognize a problem, it did not exist."[42]

In the end, the most important challenge to Minnesota's eugenics program came not from the politicians or professionals but from the so-called feebleminded who had the most at stake. Lawsuits seeking restoration to capacity appear to have increased in the late 1930s and early 1940s, but one dramatic case sent shock waves through the state bureaucracy. Rose Masters, a Catholic farm wife and mother of ten, took her case all the way to the Minnesota Supreme Court—and won. Mrs. Masters and her husband were poor but respectable tenant farmers who married in 1923 and had a large family of "apparently normal children," but as the Depression continued and their family grew, they wound up on county relief. Beginning in 1937, they received regular visits from agents of the county welfare board, who grew increasingly frustrated with Mrs. Masters's poor housekeeping and the fact that she kept having babies despite being on the dole. Several months after the birth of her ninth child, the welfare board made its move. Mrs. Masters was committed for the purpose of sterilization in 1940 and sent to the state institution at Faribault two years later, when her tenth baby was seven months old. Three months after that, her neighbors petitioned for her release. The district court rejected their appeal, but its decision was reversed by the Minnesota Supreme Court.[43]

The *Masters* case focused on the involuntary institutionalization and mental capacity of the Catholic farm wife. Sterilization was not mentioned in the court records, no doubt because the state could not act without the family's consent, and neighbors took action against commitment only after Mrs. Masters was institutionalized. Still, the case is a powerful example of the punitive welfare function of Minnesota's eugenics program and a useful illustration of three very different perspectives on welfare, feeblemindedness, and mothering at a time when the relationships between families and governments were changing. On one side, testifying for the state, was the control board psychologist who tested Mrs. Masters. He saw her IQ score of 64 as incontrovertible proof that she was a moron and could never be cured. In testimony that the Supreme Court found decidedly unconvincing, the psychologist dismissed all social or behavioral indicators of intelligence; in his view, IQ was all that mattered. On a second side, also supporting the state, was the social worker from the county welfare board. She emphasized social considerations—Mrs. Masters's repeated pregnancies, welfare dependency, disorderly home, and unlovely children—as the starkest evidence of the woman's mental deficiency. The large family squeezed into a dilapidated five-room house, tattered clothes were piled in the corners, chickens roamed through the house, and the children were often absent from school. To the social worker, Mrs. Masters's welfare dependency, disorderly household, and unkempt children proved she was a moron; nothing else mattered. On the third side of the triangle were the neighbors who opposed the action of the state. They insisted that Mrs. Masters was a "perfectly capable" mother, despite her messy house, and the whole family was normal. To the neighbors, it was a "shame for a mother with small children to be taken from them like that"; Mrs. Masters must have been institutionalized simply as "punishment for having ten children." The Supreme Court found their viewpoint understandable but incorrect. "Even in this modern age of birth control and social welfare agencies, the circumstance of being the mother of an unusually large family . . . should not label a woman as a moron," the court decreed, but it was not the number of children but their mother's inability to care for them that was at issue. Even if Mrs. Masters were restored to capacity, she was not necessarily fit to care for her children. They could remain wards of the state.[44]

In reversing the commitment order of the lower court, the Minnesota Supreme Court issued an unequivocal repudiation of eugenics and of the broad designation of mental deficiency that led to large numbers of commitments and sterilizations in the 1930s. The court rejected the psychologist's assertion that Mrs. Masters had a permanent, inherited birth defect that could be diagnosed with certainty from an IQ test alone. "The statement frequently made that all persons with IQ's below 70 are feeble-minded is not justified. . . . Intelligence is made up of too many factors to permit of such a dogmatic statement," the higher court declared. "Feeble-mindedness, viewed from a sociolegal rather than a purely medical standpoint, is not necessarily a 'permanent' and 'incurable' condition." Mrs. Masters's strong performance and "intelligent responses" on the stand constituted "persuasive proof" that, "having been given the chance of complete physical and emotional convalescence at the state institution for more than a year, her condition has definitely improved." While the *Masters* case was a significant victory for welfare-dependent families forcibly committed to state guardianship in the 1930s, it is clear that she was restored to capacity only because she was a long-standing member of the community, had respectable neighbors and a local doctor willing to support her, and demonstrated her intelligence in court. Like thousands of other Minnesotans during the Depression, Mrs. Masters was welfare-dependent, but she was not mentally retarded. The court's repudiation of her feebleminded classification tells us as much about changing attitudes toward welfare as it does about feeblemindedness and eugenics.[45]

As the *Masters* case was moving through the courts, Minnesota's sterilization program was winding down. A wartime staff shortage at the Faribault Hospital made routine tubal ligations impossible, and a robust economy created new employment opportunities for "retarded" citizens outside the state institutions while weakening public fears of the feebleminded menace. Faribault officials reported only 33 sterilizations in 1944, fewer than one-fifth of operations at the program's peak. Yet sterilization continued to be a political issue. In 1946, charges of mass sterilization at the state institution became front-page news in a bitter primary election campaign. Ironically, this time it was former Farmer-Labor governor Hjalmar Petersen, now running as a Republican, who accused the Republicans in government of mistreating inmates and sterilizing them against their will. A grand jury was called to investigate the charges, but found the complaints of "wholesale and unauthorized sterilization . . . unwarranted and not substantiated." Sterilizations were performed at Faribault under the Republican administration, as they had been under the Farmer-Laborites, but they were sanctioned by law.[46]

Thomson testified before the grand jury, but her attention was increasingly focused on parent advocacy. By encouraging parent activists to start their own organization, the National Association for Retarded Children, instead of joining the professionally oriented American Association for Mental Deficiency, she acted as midwife to the parent reform movement. Years later, mental health reformer Gunnar Dybwad praised Thomson's "enlightened concern for the retarded that was far ahead of her time." In 1959, the year she retired, a young reform-minded psychiatrist, David J. Vail, became director of medical services at the Minnesota Department of Welfare and launched a frontal assault on routine sterilization and other dehumanizing practices in the state institutions. Minnesota's eugenics era had come to a close.[47]

Or had it? The Kennedy-Johnson era brought "hope for the mentally retarded" and an antipoverty campaign that greatly expanded welfare rights, but old eugenics habits persisted when it came to poor people thought to have a mental disability. In 1965, two very different works about mentally retarded Minnesotans, both written by professors at the University of Minnesota, highlighted the coexistence of old and new ideas about eugenics, mental deficiency, and welfare. *Mental Retardation: A Family Study* (1965), a 700-page tome by geneticists Sheldon and Elizabeth Reed, followed up on the eugenics field research started in the 1910s by A. C. Rogers and the Eugenic Record Office. The Reeds, who were affiliated with the university's Dight Institute for Human Genetics (where Sheldon was director), studied the descendants of 549 patients from the Faribault School, eventually compiling data on more than 82,000 individuals and concluding—not unlike their eugenicist forebears—that mental retardation ran in families. Whether the cause of mental retardation was primarily genetic or environmental, they observed, the most significant predisposing factor was a mentally retarded relative. The solution: reduce the fertility of the unfit. "When voluntary sterilization becomes a part of the culture of the United States, we should expect a decrease of about 50 per cent per generation in the number of retarded persons."[48]

University of Minnesota law professor Robert J. Levy took a very different perspective. In a *Minnesota Law Review* article acerbically titled "Protecting the Mentally Retarded," Levy delivered a blistering attack on Minnesota's much-praised guardianship program. Condemning the gap between the rhetoric of protection and routine abuse of power, he reported that procedural safeguards were "almost entirely ignored," and he accused county welfare departments and probate judges of manipulating the guardianship program: they made commitments to reduce county costs, institutionalized delinquents who were not retarded, and pressured state wards to consent to sterilization. "Welfare departments have misused the guardianship program," Levy charged, but "many probate judges have completely abdicated their judicial responsibilities." Faced with growing pressure from reform-minded insiders like Vail, as well as from parents, former patients, and outside advocates like Levy, the state enacted significant legal changes. The 1975 Mental Retardation Protection Act gave Minnesotans with developmental disabilities greater protection from unwanted sterilization, and in 1998 the Faribault Regional Center was closed.[49] Yet the basic concerns that drove Minnesota's eugenic sterilization program in the 1930s—poverty, fiscal politics, and fear of dependency and disorder—continue to shape social welfare policy. Welfare programs, like Temporary Assistance for Needy Families (TANF), are still characterized by local variation, political manipulation, and dubious questions about maternal fitness, and local trial courts still issue sentencing decisions that curtail the parenting rights of women who would have been labeled feebleminded in another era. Looking back at the welfare function of eugenic sterilization in the 1930s, perhaps we can better see the eugenic function of welfare programs today.[50]

NOTES

1. In this essay, I use problematic terms such as *feebleminded, defective,* and *mentally retarded* freely, usually without quotation marks. These words are recognized as highly offensive today, but that is precisely the point: they capture the perceptions of their time. They are also more inclusive than current terminology, like developmental disability, because they refer to people and attributes now seen through a cultural or psychological lens.

2. Mildred Thomson, "Supervision of the Feeble-Minded by County Welfare Boards," paper given to the American Association of Mental Deficiency, May 1940, Records of the Minnesota Department of Public Welfare, Psychological Services Bureau (DPW-PSB), Minnesota Historical Society, St. Paul (MHS).

3. E. J. Engberg to Carl Swanson, June 22, 1946, Superintendent's Correspondence, Records of the Faribault State School and Hospital (FSSH), MHS.

4. Harry Bruinius, *Better for All the World: The Secret History of Forced Sterilization and America's Quest for Racial Purity* (New York: Vintage Books, 2007).

5. Some 1,930 sterilizations, or 82 percent of the total, were performed on persons classed as mentally deficient. "Sterilizations Performed under U.S. State Sterilization Statutes through December 31, 1963," February 1964, Association for Voluntary Sterilization, Supplement, Social Welfare History Archives, University of Minnesota; Philip R. Reilly, *The Surgical Solution: A History of Involuntary Sterilization in the United States* (Baltimore: Johns Hopkins University Press, 1991), 140–43.

6. Gunnar Broberg and Nils Roll-Hansen, *Eugenics and the Welfare State: Sterilization Policy in Denmark, Sweden, Norway, and Finland* (East Lansing: Michigan State University Press, 1996). See, however, Johanna Schoen, *Choice & Coercion: Birth Control, Sterilization, and Abortion in Public Health and Welfare* (Chapel Hill: University of North Carolina Press, 2005).

7. Cheuh-Fang Ma, *One Hundred Years of Public Services for Children in Minnesota* (Chicago: University of Chicago Press, 1948), 111. See also Molly Ladd-Taylor, "The 'Sociological Advantages' of Sterilization: Fiscal Politics and Feeble-Minded Women in

Interwar Minnesota," in *Mental Retardation in America: A Historical Anthology,* ed. Steven Noll and James W. Trent (New York: New York University Press, 2004), 281–99; Gary Phelps, "The Eugenics Crusade of Charles Dight," *Minnesota History* 49 (Fall 1984): 99–108. On public welfare generally, see Michael B. Katz, *In the Shadow of the Poorhouse: A Social History of Welfare in America* (New York: Basic Books, 1986).

8. Ma, *One Hundred Years,* 90; Edward MacGaffey, "A Pattern for Progress: The Minnesota Children's Code," *Minnesota History* 41 (Spring 1969): 229–36. See also Molly Ladd-Taylor, "Coping with a 'Public Menace': Eugenic Sterilization in Minnesota," *Minnesota History* 59 (Summer 2005): 237–48.

9. Mildred Thomson, *Dr. Arthur C. Rogers: Pioneer Leader in Minnesota's Program for the Mentally Retarded* (Minneapolis: Minnesota Association for Retarded Children, n.d.), MHS.

10. Mildred Thomson, *Prologue: A Minnesota Story of Mental Retardation Showing Changing Attitudes and Philosophies* (Minneapolis: Gilbert, 1963), 57.

11. *Minnesota Statutes,* Laws 1925, Chap. 154; "Sterilizations through 1963."

12. Esther Benson, "Organization of Public Welfare Activities in Minnesota" (M.A. thesis, University of Minnesota, 1941), 62.

13. Charles F. Dight, *History of the Early Stages of the Organized Eugenics Movement* (Minneapolis: Minnesota Eugenics Society, 1935), 9–10; "Human Thoroughbreds— Why Not?" pamphlet, 1922, Charles F. Dight Papers, MHS; Neal Ross Holtan, "The Eitels and Their Hospital," *Minnesota Medicine* 86 (September 2003): 52–54.

14. Edwin Black, *War against the Weak: Eugenics and America's Campaign to Create a Master Race* (New York: Four Walls Eight Windows, 2003), xv, 7.

15. "Human Thoroughbreds," 32; Charles. F. Dight, "The American Beet Sugar Company," typescript (c. 1925) and "Relating to Minnesota Eugenics Society," pamphlet, n.d., Dight Papers. See also Neal Ross Holtan, "Breeding to Brains: Eugenics, Physicians, and Politics in Minnesota in the 1920s" (M.A. thesis, University of Minnesota, 2000); Mark Soderstrom, "Weeds in Linnaeus's Garden: Science and Segregation, Eugenics, and the Rhetoric of Racism and the University of Minnesota and the Big Ten, 1900–1945" (Ph.D. thesis, University of Minnesota, 2004).

16. Charles F. Dight, "Facts Relating to My Efforts to Secure Enactment of an Eugenics Sterilization Law in Minnesota in 1931," typescript, Dight Papers; Thomson, *Prologue,* 55.

17. "History of the Sterilization of the Feeble-Minded," typescript, n.d. [1939], Superintendent's Correspondence, FSSH.

18. *Minnesota Statutes,* Laws 1917, chap. 344; Agnes Crowley, "New Laws Relating to the Feeble-Minded and Some Statistics," *Proceedings of the Minnesota State Conference of Social Work* (September 22–25, 1923), 268–72.

19. *Minnesota Statutes,* Laws 1917, chap. 344.

20. See A. F. Tredgold, *A Textbook of Mental Deficiency* (New York: William Wood, 1908); "Examination of Dr. A. W. Nuetzman," transcript in "Petition for Restoration to Capacity of Rose Masters," O. M. Teubner and Lee Adams, Appellants vs. State of Minnesota, Martin County (17th Judicial District), Case Files #33719, Minnesota Supreme Court, MHS.

21. Mildred Thomson, "My Thirty-five Years of Work with the Mentally Retarded in Minnesota," typescript, 1961–62, 133, MHS; "Report on Census of the Feeble-Minded," typescript, n.d. [1936], Dept. of Public Welfare Library, MHS. On feebleminded-

ness, see James W. Trent Jr., *Inventing the Feeble Mind: A History of Mental Retardation in the United States* (Berkeley: University of California Press, 1994). On the culture of poverty, see Alice O'Connor, *Poverty Knowledge: Social Science, Social Policy, and the Poor in Twentieth-Century U.S. History* (Princeton, N.J.: Princeton University Press), 99–123.

22. Mildred Thomson, "Social Aspects of Minnesota's Program for the Feeble-minded," *Proceedings from the American Association on Mental Deficiency* 44 (1939–40): 238–43; Raymond Koch, "The Development of Public Relief Programs in Minnesota, 1929–1941" (Ph.D. diss., University of Minnesota, 1967).

23. Thomson, *Prologue*, 6–13, 55–59, 95; "History of the Sterilization"; Mildred Thomson, Memo to Members of the Child Welfare Boards of Minnesota, n.d., Gilman (Robbins and Family) Papers, MHS.

24. Thomson, *Prologue*, 170–71; "A History of the National Association for Retarded Children, Inc.," http://www.thearc.org/NetCommunity/Page.aspx?pid=272.

25. Reilly, *Surgical Solution*, 118–22; Phelps, "Eugenics Crusade," 108. See Sharon M. Leon, "'A Human Being, and Not a Mere Social Factor': Catholic Strategies for Dealing with Sterilization Statutes in the 1920s," *Church History* 73 (June 2004): 383–411; D. Jerome Tweton, *Depression: Minnesota in the Thirties* (Fargo: North Dakota State University, Institute for Regional Studies, 1981).

26. Reilly, *Surgical Solution*, 77–78; Diane B. Paul, *Controlling Human Heredity: 1865 to the Present* (Atlantic Highlands, N.J.: Humanities Press, 1995), 84–85; Daniel J. Kevles, *In the Name of Eugenics: Genetics and the Uses of Human Heredity*, rev. ed. (Cambridge, Mass.: Harvard University Press, 1995), 114–15.

27. Engberg to Swanson. June 22, 1946.

28. Koch, "Public Relief Programs"; Ma, *One Hundred Years*, 225–35; Thomson, *Prologue*, 101.

29. Richard Lowitt and Maurine Beasley, eds., *One Third of a Nation: Lorena Hickock Reports on the Great Depression* (Urbana: University of Illinois Press, 1981), 134–35; Thomson, *Prologue*, 101, 109.

30. Thomson, *Prologue*, 109, 91–98; Mildred Thomson, "One of the most serious welfare problems . . ." untitled typescript (c. 1937), DPW; Alice Kessler-Harris, *In Pursuit of Equity: Women, Men, and the Quest for Economic Citizenship in Twentieth-Century America* (New York: Oxford University Press, 2002), 3–18; Elizabeth Faue, *Community of Suffering and Struggle: Women, Men, and the Labor Movement in Minneapolis, 1915–1945* (Chapel Hill: University of North Carolina Press, 1991). See also Deborah A. Stone, *The Disabled State* (Philadelphia: Temple University Press, 1984).

31. Thomson, "Social Aspects"; Thomson, *Prologue*, 79–80.

32. Record of Sterilization Cases, 1916–37, FSSH; Elizabeth Piper, "Analysis of Faribault Sterilization Records" (2003), unpublished report in author's possession.

33. Mildred Thomson, "To Child Welfare Board Members and Others Concerned with Carrying Out the Sterilization Law," December 12, 1934, DPW.

34. Stella Hanson, "Outside Supervision of the Feebleminded," c. 1940, DPW-PSB.

35. Charles E. Dow, "The Problem of the Feeble-Minded III," typescript, August 9, 1934, DPW Library; E. J. Engberg, "The Treatment of Mental Defectives in Minnesota," *Minnesota Medicine* 23 (May 1940), reprint, FSSH.

36. Meridel Le Sueur, "Sequel to Love," *Anvil* 9 (January/February 1935): 3–4.

37. Minnesota Legislature, Joint Committee on Acts and Activities of the Various Governmental Departments of the State of Minnesota, *Report from the Joint Senate and House Investigating Committee Covering the Acts and Activities of the Various Governmental Departments and Agencies of the State of Minnesota,* A. O. Sletvold, chairman, 1939–40 interim, MHS; A. L. Haynes, memorandum re: Miss Blanche Harkner, June 19, 1939, FSSH; Engberg to Swanson, June 22, 1946.

38. Carl Swanson to E. J. Engberg, April 12, 1940, FSSH; United States Public Health Service, Section X of a Survey of the State Hospitals of Minnesota, Section I, Washington, D.C., 1939, typescript, DPW-PSB; "From Supplementary Report on the Mental Health Program in Minnesota by the American Public Welfare Association," typescript, n.d., DPW-PSB; Thomson, *Prologue,* 125–30. See Gerald Grob, *Mental Illness and American Society, 1875–1940* (Princeton, N.J.: Princeton University Press, 1983).

39. "Supplementary Report," 4.

40. "Supplementary Report," 2.

41. "Survey of the State Hospitals of Minnesota."

42. Thomson, *Prologue,* 127–29; Mildred Thomson to Walter W. Finke, "Report of the American Public Welfare Association," memorandum, January 7, 1941, DPW-PSB.

43. *In re Masters, Teubner et al. v. State,* Supreme Court of Minnesota, March 3, 1944, 13 NW 2d, 487–94; "Petition for Restoration to Capacity of Rose Masters," O. M. Teubner and Lee Adams, Appellants vs. State of Minnesota, Martin County (17th Judicial District), Case Files #33719, Minnesota Supreme Court, MHS; Martin County Welfare Board, Minutes, December 8, 1938, MHS.

44. Testimony from "Petition for Restoration to Capacity of Rose Masters"; *In re Masters,* 487–89.

45. *In re Masters,* 490–93.

46. Thomson, *Prologue,* 147–48; Engberg to Swanson, June 22, 1946; "Cruelty in State School Charged," *Veterans' News* (June 19, 1946), MHS.

47. Gunnar Dybwad, "From Feeblemindedness to Self-Advocacy: A Half Century of Growth and Self-Fulfillment," paper presented at the Annual Meeting of the American Association of Mental Retardation, June 2, 1994, www.mncdd.org/parallels2/pdf/98-FFA-HSU.pdf. See the excellent website of the Minnesota Governor's Council on Developmental Disabilities, "Parallels in Time: A History of Developmental Disabilities," http://www.mncdd.org/parallels/index.html.

48. Elizabeth W. Reed and Sheldon C. Reed, *Mental Retardation: A Family Study* (Philadelphia: W. B. Saunders, 1965), 77–78. Many academic reviewers expressed concern about the Reeds' subjective definition of mental retardation and the validity of their data, and in 1985 the university distanced itself from eugenics by dropping "Dight" from the institute's name. See the review by Edgar A. Doll in the *American Journal of Psychology* 79 (June 1966): 346–47, and "Institute May Be Renamed Due to Benefactor's Beliefs," *Minnesota Daily* (May 2, 1985). See also Molly Ladd-Taylor, "'A Kind of Genetic Social Work': Sheldon Reed and the Origins of Genetic Counseling," in *Women, Health, and Nation: Canada and the United States since 1945,* ed. Gina Feldberg, Molly Ladd-Taylor, Alison Li, and Kathryn McPherson (Montreal: McGill-Queens University Press, 2003), 67–83.

49. Robert J. Levy, "Protecting the Mentally Retarded: An Empirical Survey and Evaluation of the Establishment of Guardianship in Minnesota," *Minnesota Law Review*

49 (1965): 821–87, quotation on 877; *Minnesota Statutes*, Laws 1975, chapter 252A. Conditions within the institutions were challenged by a 1972 lawsuit, *Welsch v. Likins*, which led to a systemwide consent decree and contributed to the closure of the state institutions. See "Parallels in Time" and Joseph P. Shapiro, *No Pity: People with Disabilities Forging a New Civil Rights Movement* (New York: Three Rivers Press, 1994).

50. See Dorothy Roberts, *Shattered Bonds: The Color of Child Welfare* (New York: Basic Books, 2001) and *Killing the Black Body: Race, Reproduction, and the Meaning of Liberty* (New York: Pantheon, 1997); Rachel Roth, "'No New Babies?' Gender Inequality and Reproductive Control in the Criminal Justice and Prison Systems," *Journal of Gender, Social Policy, and the Law* 12 (2005): 391–425.

SEVEN

Reassessing Eugenic Sterilization: The Case of North Carolina

JOHANNA SCHOEN

On March 5, 1968, Elaine Riddick, a 14-year-old African American girl from Winfall, North Carolina, was sterilized under authority of the North Carolina Eugenics Board. Elaine had just given birth to a baby boy—after being repeatedly raped by a 20-year-old man with a history of assault and incarceration. Taught not to talk about sex and fearing for her life as her rapist had threatened to kill her if she reported him, Elaine did not tell anybody about those rapes. Both of Elaine's parents were alcoholics who were intermittently in prison. When her father lost custody of his children, Elaine's six younger siblings were placed in an orphanage while Elaine and her older sister went to live with their grandmother, Miss Peaches. After Elaine's pregnancy became apparent, a Perquimans County social worker concluded that Elaine must be both promiscuous and feebleminded and drew up a petition for eugenic sterilization. Although he no longer had custody of Elaine, her father was allowed to sign the consent form and, after authorization by the Eugenics Board, Elaine was sterilized.[1]

Elaine was one of more than 7,000 people sterilized between 1929 and 1975 under the authority of the North Carolina Eugenics Board. Eugenic science gained the ear of policy makers in the last quarter of the nineteenth century when many feared that immigration and the development of birth control threatened a strong United States dominated by a native-born white population. During the 1910s and 1920s, eugenicists helped to shape legislation that aimed to stem this perceived threat by restricting marriage, controlling immigration, and sterilizing members of the community whose offspring they considered undesirable. A U.S.

Supreme Court decision in 1927, *Buck v. Bell,* upholding a Virginia court order to sterilize 17-year-old Carrie Buck for her supposed feeblemindedness, encouraged promoters of sterilization programs throughout the United States. The court was persuaded not only that Carrie Buck and her mother were "feebleminded" but also that Carrie Buck's 7-month-old daughter had inherited the family's feeblemindedness. Henceforth it was within the power of any state, unless specifically forbidden by its own constitution, to enact sterilization legislation. The Supreme Court decision was followed by a wave of new sterilization laws, and by 1929, 30 states—including North Carolina—had passed sterilization laws inspired by eugenic science.[2] From the passage of the first sterilization law until the mid-1970s, when the last states ceased operation of their sterilization programs, over 63,000 people nationwide received eugenic sterilizations. With almost 20,000 sterilizations between 1909 and 1953, California led the way, followed by North Carolina and Virginia with over 7,000 sterilizations between the late 1920s and the mid-1970s.[3]

The wave of sterilization legislation of the 1910s and 1920s mainly addressed what eugenicists termed "inheritable feeblemindedness." During the first decades of the twentieth century, eugenic theorists thought that feeblemindedness was the result of defective "germ plasm." This germ plasm, eugenicists believed, was carried from generation to generation. As the existence of bad germ plasm could not be diagnosed medically, one could recognize feeblemindedness only by such social symptoms as poverty, promiscuity, criminality, alcoholism, and illegitimacy—phenomena that were considered to lie at the root of many societal ills. Since eugenic scientists considered feeblemindedness, and thus undesirable social behaviors, to be hereditary, sterilization seemed to offer an easy medical solution to complex social problems.[4]

Thanks to the development of intelligence testing in the 1910s, scientists felt confident that they could measure mental capacity and confirm the diagnosis of feeblemindedness to which certain social problems had pointed. To eugenicists, tests of the Intelligence Quotient (IQ) were an unfailing tool for the diagnosis of feeblemindedness, lending scientific legitimacy and authority.[5] Testing allowed for the implementation of sterilization programs that depended on the ability of authorities to verify a diagnosis of feeblemindedness. An IQ rating of 70 and below quickly

identified the feebleminded and permitted health and welfare professionals to move from diagnosis to treatment.

North Carolina's eugenic sterilization law fit squarely with state
public welfare policies. Introduced by a former member of the Burke
County Board of Public Welfare, the law permitted the sterilization of
individuals who were "mentally diseased, feeble minded, or epileptic"
and whose sterilization was considered "in the interest of the mental,
moral, or physical improvement of the patient or inmate or for the public
good."[6] Between 1929 and 1932, state officials made somewhat halfhearted
use of the law, sterilizing only 49 people. But in 1932, a lawsuit resulted
in the redrafting of the state sterilization statute and the formalization
of sterilization procedures. The new law, introduced by a member of the
board of directors of Caswell Training School, established a state eugenics board composed of the commissioner of public welfare, the secretary
of health, the chief medical officers of the state hospital at Raleigh and of
an institution of the feebleminded or insane, and the state attorney general. Housed under the Department of Public Welfare, the board received
petitions for sterilization from the state's penal and charitable institutions
or county superintendents of public welfare and voted on the authorization of these petitions.[7] North Carolina was the only state in the nation
to extend the power of filing sterilization petitions to social workers; its
eugenic sterilization program represented more clearly than any other the
state's interest in sterilization.

IMPLEMENTING EUGENIC STERILIZATION: CONSTRUCTING THE FEEBLEMINDED

With the passage of the state's eugenic sterilization law, health and welfare officials implementing the program brought their own policy goals
to the table. These goals could range from controlling welfare spending
to improving the health of sterilization candidates to easing institutional
overcrowding by sterilizing and then releasing inmates from the state's
training schools. Having identified an individual as in need of eugenic
sterilization, petitioners put together a sterilization petition containing
information on the client's social, medical, and eugenic history and submitted the application to the North Carolina Eugenics Board. At their

monthly board meetings, Eugenics Board members reviewed these petitions and voted on the sterilization decision. In 95 percent of the cases, the board authorized the sterilization. Once it did, the case was assigned to the closest hospital where staff surgeons performed the surgery.

Health and welfare authorities hoped that eugenic sterilization could aid in the fight against social ills. Social and economic factors took center stage in the implementation of eugenic sterilization. Class background played an important role in identifying the mentally defective. Eugenic scientists emphasized the inability of the feebleminded to compete economically with others around them. Feebleminded persons, one official definition read, might be "capable of earning a living under favorable circumstances, but are incapable . . . of competing on equal terms with their normal fellows."[8] Poor, rural families were particularly likely to be perceived as feebleminded. By the 1930s, as the United States was struggling through the Great Depression, health and welfare officials across the country conflated welfare dependency with feeblemindedness and suggested that sterilization might provide a solution. A host of local studies seemed to confirm a close link between poverty, rural isolation, and feeblemindedness. As a 1937 *Study of Mental Health in North Carolina* reported:

> Scattered rural communities in poor "marginal" lands with low standards of living and low levels of cultural-intellectual development show high frequencies of apparent mental deficiency.[9]

If poverty pointed to the existence of feeblemindedness, receiving financial aid from the state provided the state with a financial interest in eugenic sterilization and further helped officials to identify potential sterilization candidates. A study of mental illness, mental deficiency, and epilepsy conducted in the late 1940s found that most of those classified as mentally deficient had received some type of financial assistance from the state.

> This has been in the form of general assistance, old-age assistance, ADC [Aid to Dependent Children], boarding home placement, institutional care in homes for children, free medical care and hospitalization, and employment on the projects of the WPA [Works Progress Administration] and in Civilian Conservation Corps camps. In addition to this, there is the cost of court hearings, sentences served in the county jail, training schools, State prison, and the road camps.[10]

The inability to hold a job and a history of welfare dependency both indicated feeblemindedness and constituted a reason for sterilization. Across the country, health and welfare professionals advocated the sterilization of "charity cases."[11] The case of Mary Brewer demonstrates both welfare officials' concern with clients' economic competitiveness and their unwillingness to consider present circumstances and past employment histories in their clients' favor. Until she married, Mary Brewer had supported herself, her parents, and her 11 siblings from the age of 10 by working in a hosiery mill, a cigarette factory, and a knitting mill. She had quit work only after marriage and the birth of five children forced her to stay home. In 1932, the worst year of the Great Depression, when 28 percent of the nation's households lacked a single breadwinner, Mary Brewer was also out of work. Despite her employment history, the superintendent of public welfare in Forsyth County saw in her present inability to find paid employment a symptom of feeblemindedness. "When they reach that state [of feeblemindedness] they are rarely ever self-supporting," he argued. "They usually make their living by begging."[12]

To convince family members of the desirability of surgery, board members frequently stated that they wanted to prevent the birth of children who might be financially dependent on the family or the welfare department.[13] They inquired about families' financial resources, the kind of work they did, how much they earned, whether they had ever received relief payments or help through the WPA, and how high such relief payments had been. The state, board members emphasized, had a financial interest in the sterilization. As one board member argued, "[Patient] is a state charge and to protect herself and the state we feel that this operation would be for the best."[14] Indeed, board members claimed that the real or potential financial dependence of patients gave the state a *right* to consider sterilization and asked family members to prove that a patient "is perfectly capable of taking care of herself or of any children she might have or that any children she might have would be able to take care of themselves."[15] Such proof, of course, was impossible for anyone to render.

Patients' sexual behavior provided another indication of mental disease or deficiency. Here, too, public health and welfare professionals followed the lead of eugenic scientists who identified sexual behavior as a root cause of deviancy. One researcher who was deeply concerned about the relationship between feeblemindedness, sexuality, and deviant behavior was zoologist Charles B. Davenport, who argued that sexually immoral people were also afflicted with criminality and feeblemindedness. Prostitutes, criminals, and tramps, he said, lacked the genes that allowed modern human beings to control their primitive and antisocial instincts and thus to develop civilization.

Eugenicists worried in particular about sexually attractive yet feebleminded women. "Attractive morons abound in the community," one publication warned, illustrating the danger with photographs of two young women who were inmates of a state training school. It explained, "Girls like these, who come from defective stock yet who are trained sufficiently to pass for normal by those with superficial judgment, are the greatest menace to the race when returned to the community without the protection of sterilization."[16] A 1926 *Report of the Committee on Caswell Training School* warned that 99 percent of the "high-grade" feebleminded lived outside mental institutions and were "mixing and mingling with the general population." These feebleminded people were said to be "uncontrolled, and to a considerable extent uncontrollable."[17]

But if eugenic scientists understood sexual activity outside marriage to point to larger problems of feeblemindedness, health and welfare professionals elevated the symptom to merit treatment itself. One eugenic board member explained that the operation could be performed when "people are mentally disordered, mentally defective, or promiscuous sexually."[18] Echoing eugenicists' descriptions of the feebleminded's inability to control their primitive instincts, health and social work professionals maintained that individuals who engaged in sex outside marriage lacked the self-discipline to control their sexual urges. A 1948 study of 40 persons sterilized under North Carolina's eugenic sterilization program concluded that 22 of the 40 persons were known to be sex offenders; 19 had been diagnosed as sexually promiscuous. Other sexual delinquencies in this group included incest, "trespassing including being a peeping tom," "abduction, adultery . . . bastardy, crime against nature, indecent exposure, prostitution, seduction, using a hotel room for immoral purposes," and "uncontrolled sexual desire."[19] Of the case histories with information on patients' sexual history, 80 percent were considered promiscuous.[20] Since sexual activity outside marriage was particularly disturbing in women, they became the main target of sterilization programs. Sixty-one percent of eugenic sterilizations nationwide and 84 percent of sterilizations in North Carolina were performed on women.[21]

Social background and sexual behavior determined who came in touch with the eugenic sterilization program. Most clients of eugenic sterilization programs were poor or of marginal economic status.[22] Of those suggested for sterilization in North Carolina, the majority made their living by farming, selling tobacco and other crops, and sawing wood. One-fourth were considered unable to work and 63 percent received some form of welfare benefits.[23] Often it was the illness or desertion of the male breadwinner that plunged families into poverty. Housing conditions were poor. Residences were overcrowded, in ill repair, barely furnished, and lacking conveniences.[24] Case files from the 1960s included a couple with eight children living in a three-room shack; a four-person family living in a shack without a bed—they slept on corn shucks and cotton piled in the corner; and a family of 14 living in a dilapidated three-room house heated with a woodstove, lacking indoor plumbing, and furnished only with three beds, a couch, and a baby bed.[25]

Many patients and family members also lacked formal education. Especially among the parental generation, illiteracy was still quite common during the 1930s and 1940s.[26] Confronted with the need to make a living, many felt both intimidated by the educational system and unconvinced of the value of formal education. Parents frequently kept their children at home to help on the farm or in the household. Many children missed school because their parents lacked funds to pay for books or shoes.[27]

FROM GENES TO SOCIALIZATION: EUGENIC STERILIZATION IN THE POSTWAR ERA

While health and welfare professionals had high hopes for eugenic sterilization, the eugenic science that underlay it did not go unchallenged. Starting in the 1920s, geneticists, anthropologists, physicians, and psychologists began to engage in research that eventually undermined eugenic science's basic assumptions. With the onset of World War II, interest in eugenic sterilization began to decline. The country finally pulled out of the Great Depression, and wartime production led to full employment and a drastic reduction of the welfare rolls. The shortage of surgeons during the war caused a sharp decline in the number of sterilization operations performed, and news of sterilization abuses in Nazi Germany helped to discredit the practice. In 1942, the U.S. Supreme Court struck down an Oklahoma law that had provided for the sterilization of thrice-convicted felons. Although this decision did not overturn *Buck v. Bell,* the 1927 Supreme Court decision which had upheld Virginia's sterilization law, it did set a new precedent for judicial decisions. In most states, state-ordered sterilizations had ceased completely by the late 1940s. But this decline was countered by an expansion of sterilization programs in Georgia, North Carolina, and Virginia. In 1944, these states were responsible for sterilizing 285 patients (24 percent of the nation's total for the year). In 1958, by contrast, they sterilized 574 patients (76 percent of the nation's total).

As challenges to eugenic sterilization programs were mounting across the country, supporters embarked on a massive publicity campaign to argue for the programs' continuation and expansion. They were supported by Birthright, an organization devoted to the promotion of eugenic sterilization and the distribution of sterilization statistics, and financed by Clarence J. Gamble, heir of Procter & Gamble, who had also been instrumental in financing local birth control and sterilization clinics across the country.[28] Undisturbed by the wave of criticism from within and outside the field, a small but vocal group of health and welfare professionals across the country formed coalitions with local philanthropists, ministers, women's clubs, and professors from the local universities to lobby for the continuation of state sterilization programs. Starting in 1945, sterilization supporters initiated newspaper campaigns to explain the continuing need for sterilization to the general public.[29] "Alarming Mental Deficiency Rate Confronts State" warned the headline of the first article in a newspaper series promoting sterilization in North Carolina. The article spelled out the consequences of a high rate of mental deficiency for national defense: "14.2 out [of] 100 Men Are Rejected [for military service]."[30] With defense projects forming a major state priority, the series' author asserted the following day, the state lacked money to invest in the feebleminded and the mentally ill. Sterilization, the article suggested, was the "key in solving [the] problem of feeblemindedness in the state."[31]

Although eugenic science was discredited by the 1940s, the policy goals behind the programs remained. Eugenic sterilization programs continued both to offer a solution to social problems and to cut welfare rolls by reducing the number of children born to welfare recipients. Thus as welfare rolls grew in the 1950s and 1960s, eugenic sterilization programs in a few states expanded. Continuing such programs became particularly appealing in the postwar period, when sex outside marriage and rising illegitimacy rates seemed to threaten the stability of the American family. The "rediscovery" of poverty in the early 1960s further fueled concerns about the reproductive capacity of poor families and solidified the link between illegitimacy and innate immorality. The focus on the "culture of poverty" replaced hereditary theories as a justification for eugenic sterilization and other measures while demanding similar interventions. Almost formulaically, social workers emphasized the inadequate supervision that daughters received from their mothers and warned of the continuation of poverty and neglect from generation to generation. "Sterilization," they argued, "will prevent additional children who will never be able to realize any potential they may have because adequate care will be denied them and will restrict the third generation who are caught in this cycle of poverty and neglect."[32] By the 1950s, social policy had refashioned the theoretical foundations of eugenic sterilization to meet its purposes.

Fears about the rising cost of the Aid to Dependent Children (ADC) program led to a significant shift in the racial composition of those targeted for eugenic sterilization. While the discriminatory welfare practices of the 1930s and 1940s had excluded African Americans from ADC programs and left them largely outside social workers' sphere of influence, federal pressure and a series of new requirements relating to the implementation of ADC resulted in black women's inclusion in social service programs, bringing them into closer contact with social workers and thus with state-supported sterilization. Nationwide, the percentage of welfare recipients who were African American rose from 31 percent in 1950 to 48 percent in 1961. By the 1960s, the addition of Hispanics to the rolls produced a nonwhite majority among welfare recipients. It seemed especially pressing to save funds considering the "prevalence of illegitimacy among the lower-class Negro population" and the perception that most nonwhite unwed mothers had "no means of support except through public assistance."[33] As a result, the proportion of African Americans sterilized under the auspices of the North Carolina Eugenics Board rose from 23 percent in the 1930s and 1940s to 59 percent between 1958 and 1960 and finally to 64 percent between 1964 and 1966.[34] Overall, about 60 percent of sterilization candidates were white, and 40 percent were African American.

To intensify the fight against poverty, Ellen Winston, North Carolina's commissioner of public welfare, recommended in 1951 that the state expand its use of the eugenic sterilization program by following up on ADC families in which one family member had been sterilized to determine if other members might benefit from the surgery.[35] This new policy led not only to an increase in the number of noninstitutional sterilizations but also to a sharp rise in the number of women sterilized who had given birth to children prior to having the operation. Sixty-six percent of patients sterilized in the 1950s and 1960s had had children prior to their sterilization, and 52 percent of them had given birth outside marriage.

The new emphasis on socialization drew particular attention to the behavior of teenage girls. Social workers considered girls reared in impoverished and immoral environments to be likely to perpetuate the pattern set by their parents. As social workers began to draw up the eugenic sterilization petitions of the 1950s and 1960s, this attention to young single mothers led straight into the homes of girls such as Elaine Riddick.

WHO GOT STERILIZED?

While inmates in the state mental institutions comprised a significant portion of sterilization candidates in the early decades, with the expansion of the welfare state in the postwar period the eugenics board turned its attention to the noninstitutional poor. During the 1930s and 1940s, 60 percent of those sterilized were inmates of state mental institutions. The remaining 40 percent were noninstitutional poor. By the 1950s, psychiatrists began to express open doubt about the efficacy of eugenic sterilization, and the number of institutional petitions plummeted drastically. In its stead, the percentage of those sterilized outside institutions rose to over 70 percent in the 1950s and 1960s. Overall, 40 percent of those sterilized were inmates of state mental institutions; 60 percent were noninstitutional poor.

To be sure, eugenic board members encountered sterilization candidates who suffered from serious mental illness or retardation; 23 percent of sterilization candidates had been diagnosed with some form of mental illness. Many of them suffered from auditory and visual hallucinations, delusions, or depression. Martha B., a 29-year-old mother of four, had been diagnosed with schizophrenia. Martha claimed her children were dead, showed no interest in her newborn child, and confused her husband with her sister-in-law. She laughed and talked to herself, repeatedly tried to run away, threatened to kill family members, and attempted suicide. Social workers felt that all of these problems made her unable to give proper care to her four children. They suggested that Martha be sterilized.[36] Carla M., 32, was also diagnosed with schizophrenia. Carla was unable to do housework because she had difficulty breathing and coughed up blood. In addition, she had somatic delusions, including the belief that her body was rotting away while her brain was still alive. The superintendent of the state hospital at Morganton recommended her sterilization.[37]

Health and welfare professionals not only worried about the suffering that mental conditions might cause their patients; they also feared that patients' illnesses or retardation might lead them to physically harm their children. In fact, a number of patients did threaten their children's safety or severely neglected or abused them, sometimes causing the death of a child. Ethel W., a 40-year-old mother of seven who suffered from depression, killed her two youngest children.[38] And 21-year-old Bertha M., diagnosed with mental retardation, was the mother of a 4-month-old and a 4-year-old. After her younger child died as a result of neglect, the Department of Public Welfare explained that Bertha herself was a half-grown child who needed "someone ... responsible to care for and guide her," and the department petitioned for her sterilization.[39]

I am not suggesting that eugenic sterilization was appropriate in these cases while it might not have been in others. The above cases illustrate, however, that as far as the law was concerned, some of those sterilized did indeed suffer from severe mental illness and retardation. But cases such as the above constituted the exception in a program where most sterilization candidates were neither inmates in state mental institutions nor had threatened the lives and physical safety of their children. Most sterilization candidates came to the attention of county health or welfare officials because they or their relatives received some form of welfare benefits and officials feared that pregnancy would add yet another child to the welfare rolls.

In case files, descriptions of poverty were accompanied by observations about the sexual behavior of those suggested for sterilization. Ella Mae's social worker sought sterilization for the 29-year-old mother of four because Ella Mae "seems determined to be promiscuous."[40] Pearl's social worker argued for the sterilization of Pearl, a 21-year-old mother of six, because Pearl made "no effort to curb her sexual desires and is very promiscuous with numerous suitors."[41] Ruby, a 32-year-old single woman without children, was simply deemed "oversexed," and Rosie Lee, a 26-year-old single woman without children, was considered a "sex problem" at Caswell Training School.[42]

Most troubling, as the case of Elaine Riddick suggests, was the sterilization of rape and incest victims—the majority of whom were teenage girls.[43] The petition for Beulah, a 16-year-old white girl, noted that Beulah was pregnant. When Beulah was 12, the petition observed, her father began to rape her. The man responsible for Beulah's pregnancy was a friend of Beulah's mother who had gotten the mother's permission to have sexual relations with Beulah.[44] The social worker suggested sterilization; Beulah's father concurred and signed the petition. Leora, an 18-year-old black single girl, was also pregnant. Her petition noted that Leora "had [an] incestuous relationship with her father at age 14, but because she and her mother would not testify in court against him he was found not guilty. She is now far advanced in pregnancy and her parents refuse to attempt to establish paternity as they 'don't want to get messed up in that.'"[45] Her father had signed the consent form. And Goldie, a 13-year-old white girl, was pregnant by her brother. The social worker suggesting her sterilization wrote: "Her brother is the alleged father of her child and has been committed to training school. There is also a feeling in the community that the father has had relations with her but no proof of this has been secured. According to the brother, Goldie would make an attempt to resist him but would give in each time."[46] The parents sought sterilization and signed the consent form.

Lest anybody think that cases like these were rare, the youngest person sterilized under the eugenic sterilization program was 9 years old. There were three 10-year-olds, 31 11-year-olds, 67 12-year-olds, 116 13-year-olds, 226 14-year-olds, 335 15-year-olds, 406 16-year-olds, and so on. More than one-third of those sterilized were not even of legal age to buy a drink or vote, let alone give consent to their sterilization.

It is unclear what social workers and eugenic board members thought when they looked to eugenic sterilization as a solution to these problems. Careful attention to the words of former eugenic board members and social work professionals indicates that they were genuinely concerned with their clients and desired to help improve their lives. Several health and social work professionals expressed deep ambivalence about the program when I contacted them in the 1990s. One former eugenics board member claimed to be so haunted by his role that he refused to talk to me. Others conceded that the program sometimes carried negative associations. "It isn't something we would have volunteered to do," Jakob Koomen, a former board member, explained. "We did it because the law obligated us to." Board members questioned whether authorizing sterilizations was a function of the state. "Was this a right thing to do? Did we really have all the data at hand?" Despite such doubts, however, eugenic board members were firmly convinced that eugenic sterilization constituted an important social policy intervention. The eugenic sterilization program seemed to offer an opportunity to "make things better for . . . the flow of illegitimate children and the circumstances in which they were brought up," Koomen explained. "I think our major concern was that here was a mother who has already demonstrated to be incompetent for the raising of children . . . who was having yet another child, and who, because of her community behavior was likely to have several more. That it was an illegitimate birth was of much less concern than the fact that it was an incompetent [mother]."[47] In vivid detail, social workers recalled the grinding poverty of clients. If combined with mental illness or retardation, they explained, the situation could spell disaster.[48] Koomen concluded: "I never for a moment felt that anyone on this board was doing this as a matter of punitiveness or vindictiveness or 'this is what these people deserve.' Never, never! Or a matter of discrimination of any kind. . . . Most of us felt it was a very sad situation, felt, knew that there were hundreds and hundreds and hundreds of others just like this that never came to our attention. We were sometimes thanked for having done this—parents were glad."[49]

The impersonal nature of the process masked the fact that eugenic board members were making decisions about the most intimate aspects of sterilization candidates' lives. Since board members rarely faced sterilization candidates in person, they were assessing abilities at a distance, in the abstract. Instead, clients' abilities were expressed only in an IQ rating and brief descriptions of a client's social and economic background. The presence of some severely retarded or mentally ill individuals among sterilization candidates most likely camouflaged the fact that the majority were primarily poor and lacking education. Indeed, the social distance between sterilization candidates and eugenic board members further eased the decision making process. Board members were all heads of state agencies and of similar age and academic background. They shared a sense of purpose and competency central to the decision making process. "We would usually have a brisk discussion, think about our own background," Jakob Koomen explained about the preparation for voting on a case. Even with the best of intentions, however, social workers and eugenic board members acknowledged that the eugenic sterilization program was far from flawless. "We may well have sterilized some folk who weren't that much retarded," Koomen admitted.[50]

Systemic problems further contributed to abuse. A lack of oversight on the county level meant that some social workers overstepped their boundaries with impunity. One retired social worker recalled a colleague who sterilized his entire caseload.[51] Others misrepresented the nature of the procedure when securing consent. They claimed that sterilization was reversible or coerced clients into signing consent forms by threatening the withdrawal of welfare payments. Sometimes social workers asked relatives to sign the consent form even if those relatives lacked the legal authority to do so. Board members, too, were not above reproach. They disregarded evidence that clients had, indeed, been coerced and ignored details suggesting that sterilization candidates were merely victims of unwanted sexual attention. And they gave little thought to the fact that sterilization might further harm rape and incest victims. Moreover, those board members who were directors of state mental institutions frequently voted on petitions they themselves had submitted for their institution. They did so with the acquiescence of the state attorney general's office, which also had a member on the board. Finally, while some board members in the 1950s began to have doubts about the program, they preferred to stay away from board meetings rather than challenge the program itself. This is true both for several of the psychiatrists who sat on the board as well as for several representatives from the state attorney general's office. Indeed, in its heyday, the program was run by three individuals: typically the director of the state Board of Public Welfare, Ellen Winston, or her representative, R. Eugene Brown; R. D. Higgins from the State Board of Health; and W. R. Pierce from the State Attorney General's Office.

It took until the late 1960s for North Carolina sterilization rates to decline. Changes during that decade laid the groundwork for the dismantling of state-supported sterilization. The development of more reliable contraceptives, the onset of the civil rights and women's rights movements, and a better understanding of mental disease and deficiency all contributed to a significant shift in board members' perceptions of eugenic sterilization. In 1974, North Carolina and Virginia finally repealed their eugenic sterilization laws. That same year, the Office of Economic Opportunity formulated and distributed sterilization guidelines to ensure that patients would receive adequate counseling and be given informed consent to further safeguard the poor against the possibility of sterilization abuse.

Clearly, the history of eugenic sterilization is a history of misguided governmental policy that we do not want to see repeated. By thwarting the most intimate interests of sterilization victims in their own physical and mental health, the state did wrong as well as harm. State sterilization disappointed the most natural and realistic hopes and expectations an individual might hold: the ability to determine the size of one's family. Moreover, as victims testify to a lifetime of depression, distrust, and feelings of worthlessness as a result of sterilization, the surgery left them worse off than they had been prior to the surgery. By infringing on the victims' interest in their own health and the normal functioning of their bodies, the state violated the most important interests a person has.[52]

North Carolina has come a long way in facilitating research of its history and acknowledging responsibility for its past. Following Governor Easley's 2002 apology for the program and his appointment of a eugenics study commission, the state—through its Office of Minority Health and Health Disparities—commissioned an exhibit to remember this history. In 2009, the state erected a historical marker to call attention to the work of the Eugenics Board.[53] Last but not least, North Carolina—like Sweden and the Canadian province of Alberta before it—is considering restitution payments to its sterilization victims. These are steps of great significance, and the state should be congratulated for having come so far. But an acknowledgment of past wrongs means little unless it is accompanied by serious attempts to avoid similar mistakes in the future. We need to show both our humanity toward the sterilization victims and humility toward the policy making process. This includes the study of and education about past policy and thoughtful discussion about the principles that should guide our social policy in the future.

The ability of poor women to exercise their reproductive rights remains under constant attack. Many people continue to believe that women should not have children while they receive public assistance. As we think about these problems, we need to remember the legacy of state sterilization programs. We have to prevent family planning policies that force women to use any form of birth control against their will. And we have to ensure that all have access to education about birth control, sterilization, and abortion regardless of their race, class, age, and marital status. Finally, we have to understand that women and men have the right to decide to use or not to use such services, even if we disagree with their decision. Rights are only as strong as our willingness to tolerate the decisions of others. A full acknowledgment of the suffering of the women and men sterilized under the auspices of the North Carolina Eugenics Board must include a spirited defense of their reproductive rights.

NOTES

1. John Railey, interview with Elaine Riddick Jessie, Aug. 26, 2002. In author's possession.

2. See R. Eugene Brown, *Eugenical Sterilization in North Carolina* (Raleigh, N.C.: Eugenics Board of North Carolina, 1935); see also General Statute, chapter 35, article 7; *Biennial Reports of the Eugenics Board of North Carolina* (Raleigh, N.C.: Eugenics Board of North Carolina, June 30, 1934, to July 1, 1968). For general information on *Buck v. Bell,* see J. David Smith and K. Ray Nelson, *The Sterilization of Carrie Buck* (Far Hills, N.J.: Horizon Press, 1989); Allison C. Carey, "Gender and Compulsory Sterilization Programs in America, 1907–1950," *Journal of Historical Sociology* 11, no. 1 (March 1998): 74–105; Paul Lombardo, "Eugenic Sterilization in Virginia: Aubrey Strode and the Case of *Buck v. Bell*" (Ph.D. diss., University of Virginia, 1982); Philip R. Reilly, *The Surgical Solution: A History of Involuntary Sterilization in the United States* (Baltimore: Johns Hopkins University Press, 1991), 88, 129, 137; Moya Woodside, *Sterilization in North Carolina: A Sociological and Psychological Study* (Chapel Hill: University of North Carolina Press, 1950), 194.

3. Carey, "Gender and Compulsory Sterilization Programs"; Reilly, *The Surgical Solution*, 129, 137, 158; Wendy Kline, *Building a Better Race: Gender, Sexuality, and Eugenics from the Turn of the Century to the Baby Boom* (Berkeley: University of California Press, 2001).

4. See Paul Popenoe and Roswell Hill Johnson, *Applied Eugenics* (New York: Macmillan, 1922), 74; Henry H. Goddard, *The Kallikak Family: A Study in the Heredity of Feeblemindedness* (New York: Macmillan, 1921); Daniel J. Kevles, *In the Name of Eugenics: Genetics and the Uses of Human Heredity* (New York: Alfred A. Knopf, 1985), 46–49.

5. JoAnne Brown, *The Definition of a Profession: The Authority of Metaphor in the History of Intelligence Testing, 1890–1930* (Princeton, N.J.: Princeton University Press, 1992); Stephen Jay Gould, *The Mismeasure of Man* (New York: W. W. Norton, 1981), 146–233.

6. Brown, *Eugenical Sterilization*, 21. My ability to analyze the policy making process in North Carolina is limited by the fact that the state legislature does not keep written records of its legislative sessions. Trying to understand the motivations of state legislators who debated, formulated, and funded reproductive policy is difficult, and in the case of the passage of and amendments to the state sterilization laws, it is impossible. In fact, during most of the period under discussion, such issues were considered to be private and were not debated publicly.

7. See Brown, *Eugenical Sterilization*; General Statute, chapter 35, article 7.

8. "Definitions of Feeblemindedness," General File, box 7, file: Policies and Practices of the Eugenics Board, Social Services Record Group (SS), North Carolina State Archives (NCSA). This 1908 definition is an excerpt of the British Royal Commission on the Feebleminded and was provided for North Carolina officials concerned with identifying the feebleminded for sterilization.

9. Lloyd J. Thompson, *A Study of Mental Health in North Carolina* (Raleigh, N.C., 1937), 252–53. Also Carl N. Degler, *In Search of Human Nature: The Decline and Revival of Darwinism in American Social Thought* (New York: Oxford University Press, 1991), 40. For similar calls, see Lydia Allen DeVilbiss to Margaret Sanger, September 10, 1935, Margaret Sanger Papers, Sophia Smith Collection, reel S 10, Smith College, Northampton, N.H.

10. George H. Lawrence, "A Study Relating to Mental Illness, Mental Deficiency, and Epilepsy in a Selected Rural County" (May 1948), 14, General File, box 7, file: Woodside Study, SS-NCSA.

11. DeVilbiss to Sanger, September 10, 1935. See also Eugenics Board meetings, October 24, 1961, October 23, 1962, and October 27, 1962, Eugenics Board (EB), NCSA.

12. *Brewer v. Valk*, 204 N.C. 378 (1932), 13.

13. Hearing Case 1, Eugenics Board meeting, August 17, 1938; Hearing Case 1, Eugenics Board meeting, December 2, 1936; Hearing Cases 2 and 3, Eugenics Board meeting, April 20, 1938, EB-NCSA.

14. Hearing Case 2, Eugenics Board meeting, December 2, 1936, EB-NCSA.

15. Hearing Case 2, Eugenics Board meeting, June 15, 1938; Hearing Case 1, Eugenics Board meeting, August 19, 1936; Hearing Case 1, Eugenics Board meeting, July 21, 1937; Hearing Case 2, Eugenics Board meeting, March 16, 1938, EB-NCSA.

16. Marian S. Olden, *The ABC of Human Conservation*, publication no. 31 (Princeton, N.J.: Birthright, [1946]), 6–7, copy consulted in series 1, box 2, folder 77, Human Betterment League Papers, Southern Historical Collection (HBL-SHC).

17. *Report of the Committee on Caswell Training School in Its Relation to the Problem of the Feebleminded of the State of North Carolina* (Raleigh: Capitol Printing, 1926), 15.

18. Hearing Case, Eugenics Board meeting, January 25, 1955, EB-NCSA. It was not unusual for health and welfare professionals across the country to refer promiscuous daughters to welfare departments for sterilization. See, for instance, the recommendation by an agent of the Farm Security Administration to leave such regulation of sexuality to welfare departments. W. C. Morehead, "Outline of Talk Given to FSA Personnel in Special Rural Projects Program," September 19, 1940, Planned Parenthood Federation of America I, series 3, box 45, folder: Birth Control, California, FSA Project, 1939–1941, Margaret Sanger Papers, Sophia Smith Collection, Smith College.

19. Lawrence, "A Study Relating to Mental Illness," 7, 10. Other states mirrored these findings. Seventy-five percent of those sterilized in California, for instance, were considered "sex delinquents." E. S. Gosney and Paul Popenoe, *Sterilization for Human Betterment* (New York: Macmillan, 1930), 40.

20. In the case histories, 178 histories [22 percent] gave information on sexual behavior, and 142 of those clients were described as sexually promiscuous. In 44 petitions, patients were described as victims of rape or incest or likely to be taken advantage of. Several cases contain more than one observation.

21. Reilly, *The Surgical Solution*, 94–95, 99; Carey, "Gender and Compulsory Sterilization Programs," 84–85, 101n11; Kline, *Building a Better Race*, 53. Other states also had such a high percentage of female sterilizations. Seventy-nine percent of individuals sterilized in Minnesota between 1926 and 1946, for instance, were female. Most sterilized women in that state fell into two general categories: sexual "delinquents" or older women with a number of children on welfare. Molly Ladd-Taylor, "Saving Babies, Sterilizing Mothers," *Social Politics* 4, no. 1 (Spring 1997): 136–53, at 145, 149; Mary Bishop, "Sterilization Survivors Speak Out," *Southern Exposure* 23, no. 2 (Summer 1995): 12–17, at 14.

22. A study by eugenicist Paul Popenoe on sterilizations performed in California between 1909 and 1929 shows that "economically dependent" men and women were three times as likely to be sterilized as those who were more prosperous. Paul Popenoe, "Economic and Social Status of the Sterilized Insane," in *Collected Papers on Eugenic Sterilization in California*, ed. E. S. Gosney (Pasadena, Calif.: Human Betterment Foundation, 1930), 24. See also Ladd-Taylor, "Saving Babies, Sterilizing Mothers," 144; Bishop, "Sterilization Survivors Speak Out," 15.

23. Of 800 sterilization candidates—a sample of 10 percent of the total number of sterilization petitions—147 (18.4 percent of the sample) gave information on clients' work history; 41 (28 percent) of those households received ADC payments; 63 (43 percent) received other benefits; 39 patients (26 percent) were described as unable to work.

24. Information on the physical condition of clients' housing was available in 67 cases (8.4 percent of the overall sample). Of these, 57 (85 percent) lived in inadequate housing and 37 (55 percent) in overcrowded housing. In 14 cases (21 percent), houses were in poor repair; in 11 (16 percent), houses were poorly furnished. Six families (9 percent) lacked some or all conveniences.

25. Case 19, February 28, 1961; special meeting: Case 2, February 18, 1960; Hearing Case 1, March 22, 1966.

26. While overall rates of illiteracy had dropped to 13 percent in 1930, 27.7 percent of African Americans over the age of 21 were illiterate. For the rural farm population, white illiteracy rates stood at 10.4 percent and rates for African Americans stood at 32.6 percent. See *15th Census of the United States, 1930*.

27. See Hearing Case 2, Eugenics Board meeting, April 20, 1938, for a father's inability to pay book rent; Hearing Case 1, Eugenics Board meeting, September 21, 1938, for children who couldn't attend school because they were needed to help with the harvest; also Hearing Case 2, Eugenics Board meeting, March 16, 1938, EB-NCSA.

28. For more information on Gamble, see James Reed, *From Private Vice to Public Virtue: The Birth Control Movement and American Society*, rev. ed. (Princeton, N.J.: Princeton University Press, 1983), and Doone Williams and Greer Williams, *Every Child a Wanted Child: Clarence James Gamble, M.D., and His Work in the Birth Control Movement* (Cambridge: Harvard University Press, 1978). Also Johanna Schoen, *Choice & Coercion: Birth Control, Sterilization, and Abortion in Public Health and Welfare* (Chapel Hill: University of North Carolina Press, 2005).

29. During the late 1940s, Clarence Gamble himself published at least 20 articles in professional journals arguing for eugenic sterilization. Larson argues that Gamble did as much as any one philanthropist could have done to implement eugenic sterilization programs. Edward Larson, *Sex, Race, and Science: Eugenics in the Deep South* (Baltimore: Johns Hopkins University Press, 1995), 149–51, 155–57. See also Human Betterment League of North Carolina, "25th Anniversary, 1947—25 Years of Human Betterment—1972," series 3, box 4, folder 133 (1968–74), HBL-SHC.

30. Evangeline Davis, "Alarming Mental Deficiency Rate Confronts State," *Charlotte News*, March 27, 1945.

31. Evangeline Davis, "Crowded Caswell Simply Can't Handle Patients," *Charlotte News*, March 28, 1945.

32. Case 3, EBM, June 11, 1964, EB-NCSA. See also Case 12, EBM, June 23, 1964, and Case 1, EBM, August 25, 1964, both in EB-NCSA.

33. Pax Davis, "Sterilization Funds Urged," *Raleigh News and Observer*, August 27, 1950.

34. See minutes of the Eugenic Board meetings, EB-NCSA. No comparable data for other states is available.

35. Winston to the Superintendent of Public Welfare, December 12, 1951, box 30A, file: ADC-Eugenics Program, DSS-NCSA.

36. Case 39, EBM, May 24, 1955, EB-NCSA.

37. Case 16, EBM, November 26, 1954, EB-NCSA.

38. Case 28, EBM, September 27, 1950, EB-NCSA.

39. Case 13, EBM, March 22, 1955, EB-NCSA.

40. Case 5, June 1952.

41. Case 7, May 1964.

42. Case 6, February 1952; Case 4, April 1952.

43. A quarter of the cases providing observations about sex mentioned instances of rape or incest.

44. Case 10, November 22, 1960.

45. Case 31, June 26, 1962.

46. Case 7, July 24, 1962.

47. Jakob Koomen, interview with Johanna Schoen, May 2, 1990, Chapel Hill, N.C.

48. Murlene Wall, interview with Johanna Schoen, June 19, 1997, Charlotte, N.C.

49. Koomen interview, 1990.

50. Ibid.

51. Ed Chapin, interview with Johanna Schoen, June 19, 1997, Charlotte, N.C.

52. Joel Feinberg, *Harm to Others,* vol. 1 (New York: Oxford University Press, 1984), 3–64.

53. See North Carolina Department of Cultural Resources (http://news.ncdcr.gov/ 2009/06/18/historical-highway-marker-remembers-eugenics).

EIGHT

Protection or Control?
Women's Health, Sterilization
Abuse, and *Relf v. Weinberger*

GREGORY MICHAEL DORR

On June 13, 1973, Mrs. Minnie Relf welcomed two representatives from the Montgomery, Alabama, Community Action Council (MCAC) into her home.[1] According to Mrs. Relf, these social workers told her that they had noticed boys "hanging around" her daughters, 14-year-old Minnie Lee and 12-year-old Mary Alice. Worried that this social interaction would lead to sexual intercourse, the welfare officials escorted Mrs. Relf and the girls to a local hospital and presented Mrs. Relf with a bureaucratic solution: they placed consent forms in front of her. Mrs. Relf's illiteracy rendered her unable to read the forms; her lack of education further confused the officials' explanations. Mrs. Relf believed that signing the forms would authorize the MCAC's family planning clinic to give her daughters "some shots." Her oldest daughter, 17-year-old Katie, had been receiving injections of the then-experimental, long-term birth control drug Depo-Provera from the clinic. Assuming that she was enrolling her youngest daughters in the same program, Mrs. Relf made her mark, an uncertain X, on the signature line of each consent form. A nurse then escorted Mrs. Relf home. The next morning, Mary Alice and Minnie did indeed receive shots—of sedatives—after which they were wheeled into an operating room and surgically sterilized. Unbeknownst to Mrs. Relf, she had made her mark on surgical consent forms authorizing the sterilizations. Two weeks later, on June 27, the Southern Poverty Law Center (SPLC) filed a $1 million lawsuit on the Relfs' behalf, sparking furious protest.[2] Before the year ended, Americans learned that doctors and overzealous social workers had been targeting poor women, and especially poor women of color, in a nationwide epidemic of sterilization.

This essay examines the controversy touched off by the Relf sisters' sterilization, an event that gained national notoriety through the federal court case *Relf v. Weinberger*.[3] Often only a footnote in histories about reproductive rights, significant solely as the spark that ignited a political tinderbox, *Relf* deserves closer attention for at least three reasons.[4] First, locating *Relf* in its historical context reveals the social tensions that merged quasi-Malthusian thinking with rising political animosity toward federal welfare programs. Support for implicitly and explicitly eugenic population controls emerged from the concern that the nation faced a demographic explosion among the underclass, a "population bomb" that threatened to destroy civilization in either a hail of welfare claims or violent social revolution. Politically, the Nixon administration used population anxiety to harmonize the demands of welfare activists and the New Right on one hand, demographic alarmists and the emerging anti-abortion lobby on the other—all in the name of reelecting the president. This political opportunism had far-reaching consequences.[5]

Second, *Relf* illustrates the persistence of eugenic ideology in American social policy. State and federal efforts to curb procreation in the 1960s and 1970s devolved from earlier attempts to improve the "human stock" into operations purportedly "performed for the public benefit . . . to maintain control within state institutions and to limit welfare costs."[6] Such operations, while not meant to purify the gene pool by filtering out "undesirable" genetic traits, still amounted to eugenic interventions. Some legislators, activists, and segments of the public argued that poverty bred more poverty; preventing the poor from breeding seemed a solution. Controlling procreation both defused the population bomb and offered a panacea for a hot topic social problem: the birth of "illegitimate" children to unwed mothers receiving welfare.

Finally, the revelation of the Relfs' experience came hard on the heels of another medical fiasco involving Alabama, African Americans, and the federal government. Just a year earlier, in July 1972, journalists stunned the nation with reports of the so-called Tuskegee syphilis experiment. Americans had been horrified to learn that U.S. Public Health Service physicians had watched 399 poor African American men in Macon County, Alabama, sicken (and some die) from syphilitic infections, all the while denying them antisyphilitic therapy. Investigators and the public quickly associated the lack of informed consent in both the Tuskegee and *Relf*

incidents with the specter of Nazi medical experimentation and eugenics. The fact that both events occurred in Alabama, involved poor blacks,
and revolved around sexual function was not lost on civil rights, women's
rights, and women's health activists. Reproductive rights activists argued
that poor women should be free to elect or reject sterilization, regardless of politicians', doctors', or the public's preferences. Ultimately, this
essay explores the competing claims of individuals interested in welfare
reform, civil rights, women's health, and medical ethics. The resolution
of this contest resulted in policies that controlled poor women's access to
sterilization, then and today the favored form of long-term birth control,
in the name of protecting women's rights.

WELFARE REFORM AND THE RISE
OF POPULATION POLITICS

It is a truism to note that the early 1970s were marked by generational
tensions, changing notions of racial and gender conventions, shifting
sexual mores, and expanding reproductive rights. As the fragile civil
rights coalition shattered into identity politics, women's rights, black
power, and the antiwar movement squared off against the rising New
Right of Richard Nixon's law-and-order conservatism. The rise of "liberal" ecological and antipoverty activism, however, marched in lock step
with concern about overpopulation and a movement for Zero Population
Growth that allied political liberals and conservatives. While Rachel
Carson's *Silent Spring* and Michael Harrington's *The Other America* are
cited as influential books that shaped a generation, commentators neglect Stanford biologist Paul R. Ehrlich's 1968 best seller, *The Population
Bomb*.[7] Ehrlich's crusade to alert Americans to an impending Malthu-
sian catastrophe—overpopulation exacerbated by pollution resulting
in mass starvation and chaos—inspired a cultural mania. Popular fears
found embodiment in such disparate responses as the Nixon administration's "Birth Curb Bill" (the Family Planning Services and Population Research Act of 1970, a $382 million federal program "to control
population growth") and the 1973 Charlton Heston film *Soylent Green*
(a nightmarish vision of an overpopulated America being fed on green
wafers rendered from the corpses of euthanized people).[8] The postwar
baby and economic booms had crested in the early 1960s. By the late

1960s and early 1970s, people drew parallels between starvation and war in Indo-China and America's crowded, dirty, and decrepit inner cities that erupted in the violent "long hot summers" of the late 1960s. With the Vietnam War and the Great Society sapping the treasury, the economy sagged. Soon Americans reeled under the new economic malaise, "stagflation." Ehrlich's prognostications seemed to be coming true— overpopulation was placing unsustainable demands on scarce resources, threatening collapse.

Both the state and federal governmental response to the exploding underclass often redounded to eugenics. A number of states considered passing "punitive sterilization laws," attempting to compel poor mothers of "illegitimate" children to undergo sterilization.[9] These legislative initiatives reflected ambient attitudes in America, enunciating a view of eugenic sterilization that had been implicit since the 1910s—that the so-called feebleminded, the original targets of sterilization, actually comprised a catch-all category linked more closely to poverty and perceived antisocial behavior than to organic mental deficiency.[10] After *Relf* broke, Americans realized that coercive sterilization was not restricted to states with long-standing eugenics programs or to state bureaucrats. Instead, support for the sterilization of welfare mothers reached into private medical practices, federally funded birth control clinics, and perhaps even into the Oval Office itself.

While secondary to the Vietnam War, social unrest—and the failure of President Johnson's Great Society programs to engender tranquility— emerged as a principal issue in the 1968 elections. Urban violence had rocked the nation each summer since the Watts section of Los Angeles burned in 1965. Blamed in part on crushing inner-city poverty, the violence spurred welfare activists to demand action; they swamped welfare offices with applications for increased benefits and direct participation by recipients in policy administration.[11] As the Left argued for social meliorism and the Right advocated "law and order," population control enthusiasts claimed that unfettered reproduction among the poor, encouraged by the promise of "government handouts," fueled the social crisis. Nixon determined to quell unrest, dismantle the Great Society, and cure poverty in one stroke by linking welfare reform and population control. Combined, these measures would lead to increased opportunity and equity for all Americans.

President Nixon's version of civil rights—whether the rights and well-being of women, the poor, or racial minorities (categories that often overlapped)—revolved around economic enfranchisement. Although people were obviously unequal in their social station and abilities, Nixon's personal experience as a "self-made man" established him as an inveterate meritocrat. He told aides that "we must ensure that anyone might go to the top," if they had sufficient talent and pluck.[12] Nevertheless, holding a dismal view of women's abilities (once characterizing his wife as "excess baggage"), the president announced that he only respected women "who hold the hands of the husbands who do hold office."[13] Similarly, Nixon recorded his dim view of racial minorities: "There has never in history been an adequate black nation, and they are the only race of which this is true."[14] Unsurprisingly, the president's welfare reform and population control nostrums extended gender and racial stereotypes that penalized poor minority women the most.

For Nixon, both social prejudice and misguided welfare programs retarded individual progress—the former by erecting barriers, the latter by gutting initiative. In August 1969, Nixon announced "Welfare Reform: Shared Responsibility," aimed at alleviating destitution and providing incentives to get off the dole. The president called for replacement of the existing welfare system, job-training and placement programs, revamping the Office of Economic Opportunity, and sharing federal tax revenues with the states. Nixon hoped to replace Johnson's "War on Poverty" with this "New Federalism," supplanting government largesse with individual initiative.[15] The program's Family Assistance Plan (FAP) guaranteed an annual income and encouraged work by allowing the head of household to earn up to $720 annually without sacrificing benefits (additional earnings would result in scaled reductions until income supported the family).[16] The president "reached out to both poles of popular sentiment about welfare" to sell the FAP. Passed by the House of Representatives, protest from southern politicos, welfare rights organizations, urban welfare mothers, and the national Chamber of Commerce ultimately derailed the FAP, a revised version of which died in the Senate two years later.[17] As progressive as the FAP may look to many historians, it expired without restructuring America's welfare system.[18] President Nixon lobbied for FAP only so long as it seemed politically expedient; when it failed, he turned to population policy to reform society.

On election eve 1968, when he realized he had won, Nixon announced to his staff that it was time "to get down to the nut cutting."[19] Nixon's staffers understood his reference to imply the metaphorical castration of their political opponents and a drastic trimming of the Great Society. Before the first term was half over, however—and well before the FAP foundered—the Nixon administration was facilitating the near-literal application of this measure to the poor. Without explicitly invoking the term *eugenics,* the administration employed quasi-eugenic birth control policy in the name of population control.

Poverty and reproductive policy converged in the lives of America's minority populations, a situation magnified by Daniel Patrick Moynihan's advice to the president. Moynihan argued that only welfare policies that reinforced the "traditional" patriarchal family could restore stability and independence to black families and America's inner cities. Lacking the requisite federal money and political capital to initiate a comprehensive domestic policy, Moynihan advised that it was "better just to get rid of the things that don't work, and try to build up the few that do."[20] If the FAP failed to achieve peace, Moynihan and Nixon hoped population control programs might.

Moynihan was the conduit through which population control ideas reached Nixon. Following towering Republican population activist Nelson Rockefeller, Moynihan championed improving poor women's reproductive control to alleviate both the welfare crisis and social unrest. Birth control would be welcomed by both welfare rights activists and welfare critics. Activists sought contraceptives to liberate poor women from undesired pregnancies that strained family economies. Yet they also championed poor women's right to oppose coercive sterilization. This position presented "a double-edged sword": if women were granted complete reproductive control, society could demand that they bear only the children they could support.[21] "Illegitimate children" born to impoverished single mothers became the new social/eugenic menace. Welfare opponents promoted punitive sterilization laws targeting "irresponsible" mothers to decrease welfare expenditures. In the absence of coercive laws that would guarantee sterility, the widespread provision of birth control seemed an excellent fallback. Guaranteed to reduce births among the poor, the need for welfare, and the taxes underpinning welfare, birth control mollified the white taxpayers who formed Nixon's "Silent Majority."

Thus population policy could reconcile the Left and the Right, solidifying the GOP's hold on government.

President Nixon turned to the Office of Economic Opportunity (OEO) to administer programs aimed at poor women. Ironically, despite its nominal congruence with the president's economic enfranchisement ideology, the OEO was the Great Society program Nixon hated most and targeted for a "revamping" with the FAP. Although that overhaul failed, Nixon was able to use Great Society programs like the OEO and the Department of Health, Education, and Welfare (DHEW) to disseminate family planning services—even as he sought their dissolution as governmental agencies.

Nixon's population initiatives recurred throughout his presidency. In July 1969, the president unveiled his "birth curb" program that created a federal population office to disburse $382 million to population control groups. These funds subsidized contraceptive research, birth control counseling, and the distribution of contraceptives to the poor, first through OEO-sponsored family planning services (including the MCAC family planning clinic) and later through DHEW. In the fall of 1969, the president had DHEW create a National Center for Family Planning Services. Early in 1970, the administration almost doubled the budget for contraceptive research. On Christmas Eve 1970, the president signed the Family Planning Services and Population Research Act—the "birth curb bill"—the first such law ever passed by Congress.[22] On May 18, 1971, the OEO began funding sterilizations (but never abortions, which were illegal in most instances). Concerned that "patients be protected and provided with high quality medical care," the OEO sought "guidelines and clinical standards" to "incorporate the necessary safeguards" in promulgating sterilization. These guidelines were scheduled for release by September 1, 1971. Until that time, no OEO-funded program was supposed to provide sterilizations.[23]

The administration's increased funding and its rhetoric about patient safety were creatures of political calculus, not social conscience. Although such spending might prime the flagging economy and reduce welfare demand, it also "might help us politically, a thought that just occurred to me," the president wrote. Nixon once again used cross-cutting political pressures to nullify each other. The program's increased benefits (in the form of federally sponsored birth control) looked good to welfare activists, as did Nixon's demand that every executive department (including OEO and HEW) expend their budgets through June 1972—expenditures that temporarily increased benefits. The apparent concern for patient safeguards and the provision of only sterilization (not abortion) reassured both welfare and antiabortion activists. As the abortion debate threatened to boil, the president championed proactive family planning rather than using abortion as birth control.

Under this cover, the administration dismantled the federal antipoverty infrastructure and simultaneously curbed the reproduction of the poor, further reducing the clientele for Great Society programs.[24] Reelection secured, Nixon abandoned his support of OEO. In January 1973, the president instructed two successive OEO directors to request no congressional funding appropriation.[25] Nixon also sought to pare down DHEW, the agency that assumed responsibility for federal family planning.[26] Bemoaning both the Great Society and the failure of the FAP, the administration misdirected attention from its sterilization initiatives while eviscerating welfare programs.

Nixon's legerdemain came to light when journalists discovered that the administration had suppressed the distribution of sterilization guidelines—safeguards that would have prevented the Relfs' sterilization. In 1971, Dr. Warren Hern, chief program director for the OEO's family planning division, drafted standards that barred sterilizations on anyone lacking the legal capacity to consent and "unless the individual patient has given his [sic] informed written consent to the procedure." These stipulations would have debarred the minor Relfs. Despite "dozens of requests from Community Action Agencies all over the country for the guidelines," Hern was told that the rules "would not be issued until after the 1972 elections." He suspected political procrastination.[27] Alarmed, Hern contacted John Dean, White House legal counsel. Instead of prompting the guidelines' release, Hern's actions earned him a reprimand. He resigned in protest over the "political interference and the completely irresponsible action of the OEO in blocking the dissemination of those guidelines."[28]

After the Relfs' story broke, OEO officials attempted to distance the administration from the situation by releasing portions of Dr. Hern's guidelines and claiming they had been mailed on January 11, 1973. This move refocused attention from an apparently malfeasant executive to a federal bureaucracy and an allegedly renegade state program. Both the timing of events and the subsequent unavailability of OEO officials, however, suggest that the partial disclosure was a rear-guard attempt to cover up the guidelines' suppression. Dr. Louis Hellman, deputy assistant secretary for population affairs, then claimed that Hern's recommendations were abandoned as "discriminatory" because they "included an outright prohibition of sterilization for those covered by the department's guidelines," meaning the indigent mentally incompetent and poor children.[29] Under pressure from reporters to verify this account, however, the OEO spokesperson discovered "that all 25,000 copies of the directive were still in the Government warehouse." This revelation prompted the Relfs' lawyers to add Dean and White House aide John Ehrlichman to the list of defendants, claiming that they suppressed the guidelines. Coming the night before Senator Edward Kennedy was to begin hearings on the Relfs' experience, and amidst the revelations of Dean and Ehrlichman's role in the Watergate affair, the administration's actions took on a sinister cast.

The federal foot-dragging stemmed from the complex confluence of issues on the eve of the 1972 election. Hostility toward welfare—and particularly toward the mythic figure of the "welfare queen"—would seem to have made sterilization palatable to many Nixon supporters. This line of thinking made sterilization attractive to many SPLC supporters; letters criticizing the SPLC's involvement in *Relf* came from both political liberals and conservatives who believed poor women should have no more children than they could financially support.

Unfortunately for the administration, *Roe v. Wade* had been before the Supreme Court since December 1971, inflaming anti-abortion activism, and providing a fault line among those who favored birth control for the impoverished.[30] Ostensibly, the president had rejected abortion, although tape recordings show the impact of Moynihan's influence, as well as Nixon's racism. He argued that abortion would lead to sexual "permissiveness" and would "break the family." He then acknowledged, "There are times when an abortion is necessary. I know that. When you have a black and a white. Or a rape."[31] Ultimately, he did not want his support of sterilization linked to *Roe*, thereby becoming a political liability. The best way to keep the "birth curb" initiative *sotto voce* was to squelch the guidelines, which would have required announcement and public comment before adoption. Public hearings would provide anti-abortion and Catholic voters an opportunity to attack the administration's sterilization policy, further alienating constituencies that Nixon hoped to win.

Nonetheless, the *Relf* revelations embroiled Nixon's sterilization program in the abortion battle, welfare debates, and the Watergate debacle. Attempting to insulate the administration from *Relf,* Howard K. Phillips, Nixon acolyte and former acting director of OEO, claimed that the Montgomery Community Action Committee's action was illegal and cut off all funding to the clinic on June 28.[32] Phillips declared that politics played no part in the suppression of the sterilization guidelines. "My impression of the White House position," Phillips testified before the Senate, "was that [the guidelines] merely reflected the view of the President that federal funds should not be used for abortion or sterilization. . . . My impression was that what discomfort did exist was not with the guidelines themselves."[33] Phillips's hair-splitting effort to distinguish the method from the rules fell on deaf ears in many circles. The *Washington Post,* responsible for breaking the Watergate scandal, opined that while the administration was

> advocating less government, draconian law enforcement, more decentralization and indifference to problems of race and poverty, some of them were hiring goon squads and others were using the power of their public office for blackmail and extortion. Phillips, the ex-head of OEO, hasn't been connected with that stuff. No, he was handing out money from his agency to pay for the sterilization of black children. By comparison to Phillips, the Deans, the Colsons and the Magruders look almost good. . . . [T]his revelation may explain Richard Nixon's opposition to abortion. There was a better plan . . . the new Federalism's final solution to race and poverty in America.[34]

The Nixon administration's political machinations reflected trends established in the 1930s. Then eugenicists "became less concerned with preventing the birth of children with genetic defects and more concerned with preventing parenthood in those individuals who were thought to be unable to care for children." The goal in the 1930s, as in the early 1970s, "was to reduce new burdens on the public purse."[35] Eugenicists wanted to extend sterilization's reach beyond the institutionalized to achieve this end. The unfit, too poor or benighted to use contraceptives, might be persuaded to undergo sterilization.[36] Activists would "sell" sterilization to the unfit, offsetting the dysgenic effect of fit couples' use of contraceptives. Yet voluntary sterilization—long trammeled by so-called age/parity requirements debarring young women, mothers with fewer than three children, and women whose husbands would not also sign the consent form—remained largely unavailable until the 1970s.[37] By then, the heirs to the eugenics movement and their foes understood that education about and access to contraception represented both a boon to poor women (they could exercise greater reproductive freedom) and a doorway to potential abuse (officials could force birth control or sterilization on unwilling women).[38] Voluntarism would allow sterilization to escape the stigma of eugenics; the definition of "voluntary," however, remained fluid—conditioned by emerging concepts of informed consent. What constituted informed consent now became the fulcrum over which women's desires and societal imperatives teetered.

EUGENICS, ALABAMA, AND THE ROAD TO *RELF*

Talk of a "final solution" raised the specter of Nazi eugenics. As *Relf* broke, civil rights activist Julian Bond, then president of the SPLC, said, "Sterilization of the retarded had its precedent in Nazi Germany. This whole thing is a horrendous attack on privacy, innocence, and the right of motherhood."[39] Bond's comments amplified historical resonances that influenced consideration of *Relf* at the time and since. Bond incorrectly located the origins of eugenic sterilization in Nazi law and practice. Constitutional precedent for the eugenic sterilization of the mentally retarded, however, originated with Justice Holmes's 1927 ruling in *Buck v. Bell.* By June 1933, when the Nazis passed their eugenics laws, thousands of Americans had already undergone eugenic sterilization. Moreover, Bond's assertion that the Relfs' sterilizations violated notions of "privacy, innocence, and the right of motherhood" was itself a product of historical context. While the Supreme Court had already established that procreation represented "one of the basic civil rights of man" in *Skinner v. Oklahoma,* and had declared contraception legal and protected by the "right to privacy" in *Griswold v. Connecticut,* the then-recent decision in *Roe v. Wade* enlarged the notion of privacy to encompass bodily integrity. Neither *Skinner* nor *Roe* directly challenged eugenics or the precedent set in *Buck*.[40] Bond's remarks demonstrated that by 1973, the footprints of America's and Alabama's eugenic past had been erased by the passage of time and the redirection of attention from compulsory sterilization to voluntary birth control.

Until recently, most historians have claimed that eugenics died after the revelations of the Nazi Holocaust.[41] The *Relf* case demonstrates the durability of eugenic ideas in America generally and Alabama specifically.[42] The state of Alabama has a long history of involvement with eugenic theory and practice.[43] Although it took 19 years, Alabama's eugenics lobbyists persuaded the legislature to pass a sterilization law in 1919.[44] This statute empowered the superintendent of the Alabama Home for the Feebleminded "to sterilize any inmate" provided he had the concurrence of the superintendent of the Alabama Insane Hospitals. Since these men both worked in Tuscaloosa and were close friends, this provision amounted to a rubber stamp that gave the superintendents carte blanche, without any of *Buck*'s procedural safeguards. After 1923, the superintendencies were merged, and William D. Partlow, an avid eugenicist, assumed sole discretion over sterilization.[45]

Statistically speaking, Alabama's eugenics program was tiny compared with eugenics powerhouse California. Although Partlow sterilized every patient he discharged between 1919 and 1935, this only amounted to 224 people.[46] California, in contrast, sterilized 3,763 people in the same period.[47] Yet when Alabama eugenicists sought to extend the coverage of their eugenics laws, they looked not to California for a model but to Nazi Germany. Alabama's State Public Health Officer, Dr. James Norment Baker, noted in 1934 that "the whole civilised world will watch, with keen interest, the bold experiment just launched by Germany in mass sterilization," predicting that the Reich would realize massive savings. Lamenting the short reach of Alabama's sterilization law, Baker averred, "Both the humanitarian and economic aspects of this question are indeed so gigantic as to attract the interests of all; and especially should this whole question of human betterment via the eugenic and sterilisation route make a peculiar and lasting appeal to every physician."[48]

Testimony by Baker and others persuaded the legislature to pass a bill calling for an expansive sterilization program, larger than any other in the country. Under the proposed law, people subject to compulsory sterilization included anyone committed to state homes for the insane and feebleminded, reformatories, industrial schools, or training schools (a cipher for juvenile delinquents, the blind and deaf), as well as any "sexual pervert, Sadist, homosexualist, Masochist, Sodomist, or [people suffering from] any other grave form of sexual perversion." The bill also mandated the sterilization of "any prisoner who has twice been convicted of rape" or imprisoned three or more times. The statute would also have applied to individuals "habitually and constantly dependent upon public relief or support by charity."[49] Clearly Alabamians in the midst of the Great Depression were as concerned about lightening the economic load as they were about reducing the genetic load.

Governor Bibb Graves asked the Alabama Supreme Court for an advisory opinion on the legislation. The court found the bill unconstitutional for lack of due process provisions; Graves vetoed it. Though the 1919 sterilization law remained on the books, the advisory opinion exposed it to constitutional challenge, too. A bitter William Partlow wrote that, as a result of liability concerns, he had "positively discontinued the practice of sterilization, which will of necessity effect [sic] our previous liberal policy of granting paroles."[50] Compulsory eugenic sterilization in Alabama was not dead, however.

Despite the revelations of the Holocaust, many Alabamians remained committed to eugenics. Alabama's medical school, the principal source of the state's physicians, had trained generations of doctors in eugenics. Medical students began rotating through Partlow's facilities in Tuscaloosa when they opened in 1924, learning eugenics at the master's knee.[51] These men took eugenic ideas into their private practices. Budding social workers learned eugenics in college and came to understand the notion of the "unfit." This eugenic instruction bred convictions that led to abortive attempts to pass sterilization bills in 1945 and 1951.[52] The notion that feebleminded people *deserved* sterilization survived into the 1960s in the public mind, even though most scientists, psychiatrists, and physicians had abandoned simplistic genetic explanations for mental retardation. Nevertheless, states performed sterilizations under the aegis of their eugenics laws until 1979.[53] Federal funding provided by OEO, DHEW, and Medicaid made it easier to put thought into action. Eugenically inclined doctors were guaranteed payment for practicing their beliefs. The persistence of these ideas, and their impact on people's lives, came to light in the aftermath of *Relf.*

ABUSE UNFOLDS

Relf's filing prompted other abused women to speak out, even as official investigations revealed an ever-widening pattern of sterilization abuse. By July 2, 1973, OEO investigators had determined that 11 women, all but 1 of whom were black, had been sterilized by the MCAC.[54] On July 13, Nial Ruth Cox, a 26-year-old African American hospital worker, reported that North Carolina officials had sterilized her when she became an unwed mother at 18. Her case prompted scrutiny of the North Carolina State Eugenics Board, which routinely classified poor women as "feebleminded" and ordered them sterilized.[55] On July 21, officials from the South Carolina Department of Social Services and the Atlanta offices of DHEW began investigating physicians in Aiken, South Carolina, especially Dr. Clovis Pierce. Inspectors discovered that Pierce refused to provide obstetrical services to poor women on welfare—unless the women consented to sterilization during delivery. Rather than forgo obstetrical services, half of the welfare mothers who delivered babies in Aiken County during 1973 received sterilizations. One woman sterilized by Pierce testified that he told her "he worked hard to pay his taxes and was tired of having people come to him to have babies he would have to support with his tax dollars," despite the fact that Medicaid had paid him over $60,000 in 1972 and the first half of 1973 for serving poor women.[56] Pierce was not an ideological wildcat. A 1972 survey showed that 94 percent of obstetrician-gynecologists supported "compulsory sterilization or the withholding of welfare support for unwed mothers who already had three children."[57] The SPLC brought two of Dr. Pierce's patients into the *Relf* suit and represented them in a federal tort case, *Walker v. Pierce*.[58]

The federal government, responsible for subsidizing this cottage industry in sterilization, admitted that 16,000 women and 9,000 men had been sterilized at public expense between 1972 and 1973. Government officials, however, could not vouch for the degree of informed consent in every case. Black women had special reason for concern. Beyond the conventional wisdom about "Mississippi appendectomies" in the black community (sterilizations performed under the guise of other abdominal surgeries), independent studies demonstrated in 1970 that black women were sterilized at more than twice the rate of white women (9 per 1,000 black patients compared with 4.1 per 1,000 white patients).[59] In the wake of *Relf,* Latinas and Native American women came forward with their own stories of widespread sterilization abuse, too.[60] These revelations proved to be only the tip of America's deeply submerged sterilization iceberg.

Between 1968 and 1973, America witnessed a sterilization explosion: approximately two million Americans (one million men, one million women) underwent sterilization in 1973, up from roughly 400,000 in 1968. This increase led to additional questions. Prompted by the work of Dr. Bernard Rosenfeld, the Health Research Group (a watchdog agency funded by Ralph Nader's activist organization Public Citizen) discovered a nationwide epidemic of abuse. As an intern at Oakland's Highland General Hospital, Rosenfeld witnessed doctors "selling" sterilization to patients, particularly women of color. While helping a senior resident deliver a woman who "in his opinion, had enough children," Rosenfeld heard the doctor ask the woman "'What do you think about getting your tubes tied now?'" The physician then advised the patient's husband that he "better tell his wife to get her tubes tied," because, "'With the pill she'll get a blood clot to her brain; with the I.U.D. a pregnancy in her tubes.'" Sterilization was safer. When Rosenfeld objected to this advice, the resident responded, "'Hell, man, don't you realize we're paying for her kids?'" despite the fact that the woman was not on welfare. She just happened to be African American and delivering in the publicly supported county hospital.[61]

The Health Research Group found similar maltreatment occurring at America's best teaching hospitals. Often, when sterilization was on the line, there existed "little evidence of informed consent by the patient," and it was clear that "these operations have been 'sold' to the public by surgeons in a manner not unlike many other deceptive marketing practices." Surgeons promoted sterilization for eugenic and professional reasons. The eugenic reasons were reflected in the cases from Alabama and the Carolinas. The professional impulse stemmed from interns' and residents' need to perform a certain number of surgical procedures as a prerequisite to taking their board examinations. As a resident told interns at the University of Southern California Hospital, "I want you to ask every one of the girls if they want their tubes tied, regardless of how old they are. Remember everyone you get to get her tubes tied means two tubes for some resident or intern," moving them that much closer to certification.

The report concluded, "It is probable that of the 2 million people who undergo surgical sterilization each year, at least several hundred thousand are considerably less than well informed about the irreversibility, risks and alternative methods of family planning when the[y] 'decide' to have these operations."[62] The arena for sterilization abuse was broad, indeed.

PROTECTION OR CONTROL:
THE BATTLE OVER *RELF*

The *Relf* case broke at a critical juncture in the history of medical care and human rights. Congress had just concluded its investigation into the ethics of the Tuskegee syphilis experiment, and a federal court in Alabama was reconsidering the constitutionality of the state's moribund 1919 sterilization law in the case *Wyatt v. Aderholt*.[63] All of this occurred amidst the civil rights and women's rights turmoil of the early 1970s. Although neither media pundits nor legal officials ever linked Tuskegee, *Relf,* and *Wyatt,* exploring their commonalities is instructive. Two themes emerge from the comparison. First, in all three cases, the central issue was whether the victims—the Relf girls, the syphilitic men, and institutionalized mental patients—had given their informed consent. In *Relf* and Tuskegee, observers identified an abusive white power structure subverting informed consent by exploiting female African American healthcare workers. Digging deeper into this dynamic, however, reveals that a class-based impulse, mirroring the effects of racism, might have motivated members of an oppressed group to oppress others. The second major theme arises from the fact that *Relf* and *Wyatt* connect earlier eugenic policies to the sterilization abuses of 1973. While neither *Relf* nor *Wyatt* overturned the precedent set in *Buck v. Bell,* they both implicitly undermined the durable Progressive Era notion that, in the public interest, the state can reliably substitute its judgment for individual decision making. Analyzing these issues underscores the intended and unintended consequences of *Relf*: a ruling that protected poor women from abuse at the expense of controlling their access to birth control and hence curbing their reproductive freedom.

Immediately after the Relfs' story broke, officials started justifying their actions. Joseph Conklin, director of the MCAC, claimed that a notary public had "quizzed" Mrs. Relf, verifying that she understood the effects of sterilization, before notarizing the "signed" consent form. Conklin also asserted that the girls lacked the "mental talents" to take birth control pills and that Depo-Provera was not an option because the drug had been prohibited since May, when his program shifted birth control funding from an OEO grant to money from the DHEW.[64] Federal review called Conklin's timeline into question: DHEW investigators concluded that financial support had been transferred from OEO as of March 1, not May. Moreover, including the Relf girls, DHEW determined that the MCAC had sterilized 11 women, all but 1 of whom were black, 7 of whom were mentally retarded, and 5 of whom were minors. Although the investigators concluded that "this project is basically regarded as a good project," serving "about 2,200 of the 3,000 active family planning patients in Montgomery," the discrepancies between state and federal officials fueled public skepticism about MCAC motives and conduct.[65]

MCAC officials scrambled to deflect the insistent allegations of sexism and racism. Two days after the Relfs filed suit, Alabama State Health Department officials tried to defuse racial tensions by disclosing that 82 people, 40 of whom were white, had obtained sterilizations through health department programs—implying that, if half the people sterilized were white, the procedures must have been voluntary and nonracist.[66] Meanwhile, Mrs. Orelia Dixon, the white director of the MCAC family planning clinic, noted that she, like Mrs. Relf, was a mother of six. Mrs. Dixon implied that, as a woman and mother, she could only have had benign intentions, perhaps including a desire to spare "obviously incapable" young women from being burdened with children. Mrs. Dixon claimed that there was "no way" anyone could be sterilized without consent. "We ask them [the patients] if they understand it," Mrs. Dixon claimed. "If they say they understand it, that's it. That's about as far as you can go."[67] Mrs. Dixon argued that, since the clinic nurse/social workers who took the girls to the hospital were black (as were 11 of the clinic's 14 staff members), race could not possibly have influenced the Relfs' treatment. After all, she reasoned, African Americans would never willingly have discriminated against other African Americans. Indeed, unidentified black nurses from the clinic claimed that the Relfs "exploited the situation," and argued that other poor blacks that had been sterilized "thank God" for the clinic's services. "You can't take one isolated example like this and get a balanced picture of what the clinic does," according to Mrs. Dixon.[68]

Mrs. Dixon's assertion carried a great deal of truth, but it also illuminates the double bind created by *Relf*. According to the Relfs' lawyer, Joseph Levin, there was no question that the MCAC clinic and the operating physician believed they were acting in the best interests of their patients.[69] That said, it was the SPLC's position that "mental retardation and mental illness have been used as excuses for sterilization of poverty level people where no such necessity existed; and the practice has been widespread."[70] Although the president of the National Council of Negro Women (NCNW)—who decried the sterilization of the Relfs—urged the SPLC to support renewal of the Family Planning Services and Population Research Act as a "vitally needed health service," Levin noted that the SPLC felt the need to pursue increased protections. Intervention, even at the risk of compromising women's reproductive autonomy, was needed "to insure that children and other persons who are unable to comprehend the consequences of sterilization and experimental drugs would be protected."[71] Poor women of color deserved access to all reproductive technologies, but to ensure their freedom from abuse—and to protect minors and the mentally incompetent—they would have to clear additional hurdles before they could obtain sterilizations.

Moreover, Mrs. Dixon and other commentators ignored how, even within the community of black women, intragender and intraracial class tensions could exactly mirror the effects of sexism and racism.[72] Many African Americans adhered to a cultural eugenics that projected white eugenic judgments onto the black community. Just as whites viewed their feebleminded and immoral as eugenically "unfit," so too might black middle-class officials have considered the Relfs unfit. Just as Carrie Buck represented the "shiftless, ignorant, worthless class of anti-social whites of the South," so too the Relfs—obviously uneducated, on welfare, with young Mary Alice both physically deformed (she was born without a right hand) and apparently mentally retarded—represented the analogous segment of the black community.[73]

Comments from African American nurses in Montgomery suggest that intraracial and intragender class dynamics that operated like racism may well have been present in the Relf situation. One unnamed black nurse from the MCAC told a reporter that the lack of birth control had historically "kept back black people." How could impoverished women "provide for the education of two, much less a dozen children?" she asked rhetorically.[74] This nurse's logic elided the issue of consent in favor of patent economic considerations. Another black Montgomery County Heath Department nurse remarked, "We've had the Relfs as clients for more than five years, and for the longest time I was trying to explain that the two elder girls should have their (fallopian) tubes tied.... The family always refused." A *Jet* reporter submitted this comment as evidence that the Relfs could not have given informed consent—they had always refused in the past. Therefore, Mrs. Relf signed the consent form because her ignorance and her culturally ingrained deference to middle-class blacks and whites created a coercive situation. While this dynamic may have been operating, the *Jet* reporter neglected the undertones in the nurse's statement. The black public health nurse was *in favor* of sterilizing the elder Relf girls, Katie and Minnie Lee; she merely had more scruples than the MCAC officials (including the black nurses who visited the Relfs) and refused to hoodwink Mrs. Relf into consent.[75] That these intraracial and intragender beliefs existed may explain why, almost four months after the Relfs' story broke, no "united effort on the part of blacks ... to stop the threat has been initiated. Apparently only one white woman, Gloria Steinem, has publicly sounded the battlecry by urging that forced sterilization 'be fought and fought as a coalition issue.'"[76] The race and gender of the nurses directly involved in the Relfs' sterilization immunized their class-biased actions. Nurses in the MCAC may well have believed that the Relfs were lucky to receive sterilization and should "thank God." The Montgomery County Health Department nurse's response reflected precisely this evaluation. The appalling lack of medical care facing generations of black southerners undoubtedly made even crumbs seem like much more than half a loaf.[77]

Traditional racial mores, too, influenced these black nurses. MCAC nurses remained subordinate to white officials, including Mrs. Dixon. Even if they did not share class or racial biases against poor mothers, these women might be "on Mrs. Dixon's side" to protect their jobs. It was not exactly a case of "White folks say do, come, go, don't, and Blacks do, come, go and don't," as *Jet* magazine reported. Instead, the supposedly benign nature of these interventions may have obscured their racial overtones.[78] In the end, the "moral astigmatism" created by class bias made sterilization seem benignly eugenic, rather than eugenically genocidal.[79]

The eugenic nature of the Relf sterilizations came to light as Alabama was reconsidering, definitively, the status of the 1919 sterilization law. While that law had been in abeyance, cast into limbo by the 1935 Alabama Supreme Court advisory opinion, it remained on the books. And, despite Dr. Partlow's claims to the contrary, it appears that officials continued to sterilize residents under the aegis of the obsolete law. In *Wyatt v. Aderholt,* a federal court finally declared Alabama's 1919 sterilization statute unconstitutional. Citing the controlling authority of *Buck v. Bell,* the court noted the obvious procedural defects of the 1919 law, which left sterilization to the discretion of state officials with no provision for due process. U.S. District Judge Frank Johnson ruled that sterilizations on state wards could be performed only with their voluntary informed consent (in the case of legally competent patients), or with a medical determination that sterilization was in the best interests of the legally incompetent. Johnson mandated a due process procedure including legal representation, a judicial hearing, and direct consultation with the patient. Johnson also forbade making sterilization a condition for release, Dr. Partlow's favored tactic. With this ruling, Alabama's "official" eugenic sterilization law exceeded the standard in *Buck v. Bell* by requiring voluntary informed consent and a finding that the sterilization was "in the best interest of the resident," not just in the best interest of society or "on the basis of institutional convenience or [for] purely administrative considerations."[80] Economically motivated sterilization was now beyond the reach of the state.

That the Relf sterilizations—and the hundreds of thousands of similar federally funded procedures that were discovered by investigators—bore a eugenic cast gained notice in U.S. Federal Judge Gerhard Gesell's initial opinion in *Relf v. Weinberger.* In his decision, Gesell concluded that federally funded family planning clinics sterilized between 100,000 and 150,000 poor people per year. Although Gesell did not note it, these sterilization rates indeed matched those of the Nazi regime in the 1930s. The only difference was that informed consent accompanied *some* of the American operations.[81] Gesell then affirmed the "uncontroverted evidence" that minors, the mentally retarded, and "an indefinite number of poor people have been improperly coerced into accepting a sterilization operation."[82] Conceding that the case appeared "during a period of rapid change in the field of birth control," and that contraception and family planning had become "widely accepted," the judge cautioned that "over this entire area lies the specter of overpopulation, with its possible impact upon the food supply, interpersonal relations, privacy, and the enjoyment of our 'inalienable rights.'"[83]

Despite, or perhaps because of, the perilous nature of the situation, Judge Gesell wrote, "Surely the Federal Government must move cautiously in this area, under well-defined policies determined by Congress after full consideration of constitutional and far-reaching social implications. The line between family planning and eugenics is murky." Gesell admonished, "Whatever might be the merits of limiting irresponsible reproduction, which each year places increasing numbers of unwanted or mentally defective children into tax supported institutions, it is for Congress and not individual social workers and physicians to determine the manner in which federal funds should be used to support such a program." Gesell decried the "drift into a policy which has unfathomed implications and which permanently deprives unwilling or immature citizens of their ability to procreate without adequate legal safeguards and a legislative determination of the appropriate standards in light of the general welfare and of individual rights."[84]

Acting in the name of the inalienable individual rights that underpinned the civil rights and women's health movements, Judge Gesell struck down interim sterilization guidelines drafted by HEW and ordered the creation of new rules restricting the use of federal funds for "voluntary" sterilization. Meanwhile, he enjoined the federal government from providing funds for the sterilization of mentally or legally incompetent individuals.[85] Activists at the time worried that these restrictions might not lessen the incidence of abuse and that they might increase the obstacles poor women faced in accessing reliable birth control. It would take another four years and two more court rulings before the federal sterilization guidelines fulfilled the need for informed consent, yet still allowed maximal access to birth control services.

In November 1978, HEW issued regulations requiring a mandatory 30-day waiting period for federally funded sterilizations and the provision of translators when necessary, and banned the signing of consent forms while a patient was undergoing labor, childbirth, or abortion.[86] Informed consent protected women from abuse, but the procedures to ensure the uncoerced nature of that consent controlled women's access to sterilization. Women, it seemed to many, faced yet another double bind.[87]

UNINTENDED CONSEQUENCES

As Judge Gesell and federal officials worked to devise appropriate rules governing federally subsidized sterilizations, the Relf family awaited the outcome of their damage suit. Handled by the firm of flamboyant personal injury lawyer Melvin "the King of Torts" Belli, the suit ultimately failed because an appropriate defendant could not be identified under the Federal Tort Claims Act.[88] Joseph Levin recalled that, although the Relfs did not receive any monetary compensation, the SPLC had young Mary Alice evaluated and placed in an appropriate educational program. Her "mental retardation" was spurious—the effect not of organic mental defect but of a severe, untreated cleft palate that had made it nearly impossible for her to speak intelligibly. After arranging Mary Alice's cleft repair surgery, the SPLC lost contact with the Relf girls.[89]

Ultimately, it appears that the *Relf* decision had a limited impact in preventing sterilization abuse. In January 1975, almost nine months after Judge Gesell ordered new regulations, a survey including 42 of the nation's 50 largest teaching hospitals found 76 percent continued to violate the new federal sterilization standards.[90] A subsequent 1979 study revealed that 70 percent of hospitals remained noncompliant with the guidelines.[91] Even with the *Relf* decision in place, tort claims stemming from other coercive sterilizations also failed. Although Joseph Levin managed to win the damages suit in *Walker v. Pierce,* the jury awarded Mrs. Shirley Brown five dollars for the indignities she suffered at the hands of Dr. Pierce—a far cry from the million dollars sought by the Relfs. And Dr. Pierce won on appeal, having the initial damages decision reversed. In the final order in *Walker v. Pierce,* the federal appeals court found "no reason why Dr. Pierce could not establish and pursue the policy he has publicly and freely announced [that he would not deliver additional children to mothers on welfare unless they underwent sterilization]" as long as all patients were "made fully aware of his professional attitude." Since, in the court's opinion, Dr. Pierce's acceptance of Medicaid funds did not make him a "state agent," and there was no explicit law against his policies, he was free to continue coercive sterilization, just as his patients were "free" to find another attending obstetrician.[92]

The restriction of federal funding and a series of women's health care scandals in the 1970s and 1980s were probably more effective in curbing abusive sterilization. Lower reimbursement rates (and the prohibition against using federal funds for the sterilization of minors, mental incompetents, and the institutionalized) made the sterilization of welfare women less economically attractive to physicians who had previously profited from such procedures. Prohibiting federal funding for hysterectomies performed for birth control likely reduced both economic and training incentives for doctors engaging in gratuitous procedures. The thalidomide and D.E.S. scandals, combined with massive damage claims resulting from the Dalkon Shield IUD fiasco and the debates over the safety of silicone breast implants, brought women's health issues to the fore and placed physicians (especially obstetricians, gynecologists, and surgeons) under intense scrutiny. The concomitant trend toward massive malpractice awards against individual practitioners, and the mushrooming incidence of malpractice litigation in the 1980s, redoubled the pressure on doctors to avoid liability. Thus the antisterilization abuse enforcement regime shifted from federal regulation to private litigation. Meanwhile, poor women continued to find it difficult to exercise reproductive autonomy. As the *Washington Post* announced in 1998, "Some Poor Mothers Find Bureaucracy a Barrier to Sterilization," a situation that persists today. In the name of protecting minors, the mentally incompetent, and the poor from abuse, federal regulation of sterilization continues to control poor women's reproductive rights.[93]

NOTES

Previous iterations of this chapter were delivered at the Social Science History Association (2002), Southern Association for the History of Medicine and Science (2004), the University of Alabama–Birmingham (2005), the Suffolk Section of the Massachusetts Medical Society and Grand Rounds of the Vincent Obstetrics and Gynecology Department, Massachusetts General Hospital (2006). The author thanks participants at these meetings for helpful critiques. This narrative is distilled from newspaper reports, court documents, and the *Relf* files of the Southern Poverty Law Center. The author thanks Joseph J. Levin Jr., Center co-founder and president emeritus for granting unfettered access to the extant files.

1. "Suit Says Girls Were Sterilized," *New York Times*, June 27, 1973, 44; "Clinic Defends Sterilization of 2 Girls, 12 and 14," *New York Times*, June 28, 1973, 14; "Racism, Ethics, and Rights at Issue in Sterilization Case," *New York Times*, July 2, 1973, 10; "Relfs File Suit against U.S.," *Montgomery Advertiser*, July 19, 1973, 1; Bruce Nichols, "Suit Seeks Million," *Montgomery Advertiser*, 1; Nichols, "Sterilization Case 'Patient's Choice,'" *Montgomery Advertiser*, June 28, 1973, 1; Nichols, "Staff of Clinic Stunned by Suit,"

Montgomery Advertiser, June 29, 1973, 1; Louis Berney, "Relfs Have Scant Understanding of Happenings," *Montgomery Advertiser,* July 1, 1973, 1; Nichols, "Relfs Leave by Plane to Testify to Subcommittee," *Montgomery Advertiser,* July 10, 1973, 1; "Authorization for Sterilization" documents, Minnie Lee and Mary Alice Relf, box 2, Relf Case Files, Southern Poverty Law Center, Montgomery, Ala. (hereinafter referred to as Relf Files); "Complaint, Class Action," Civil Action 1557-73, Relf Files, 8–10; "Class Defendant Action, Class Plaintiff Action, Temporary Restraining Order Requested, Preliminary Injunction Requested," Civil Action 4099-N, Relf Files, 1–6.

2. When MCAC workers returned for Katie, she locked herself in her room. "Complaint, Class Action," Relf Files, 9–10. Social workers from the St. Jude Social Service Center, who had been assisting the Relfs, investigated the sterilizations and alerted the Southern Poverty Law Center. These actions "didn't have anything to do with Catholic beliefs," one social worker said. "It was just a personal reaction." See "The Relfs: A Family Caught Up in a National Controversy," *Washington Post,* July 8, 1973, A4; "Statement of Jessie Bly" (St. Jude's social worker) and "Chronology of Case Contacts," Relf Files.

3. *Relf v. Weinberger,* 372 F. Supp. 1196 (1974); *Relf v. Mathews,* 403 F. Supp. 1235 (1975); *Relf v. Weinberger,* 565 F. 2d 722 (1977).

4. The Relfs do not appear in landmark studies like David J. Garrow, *Liberty and Sexuality: The Right to Privacy and the Making of Roe v. Wade* (New York: Macmillan, 1994) or Linda Gordon, *Woman's Body, Woman's Right* (New York: Penguin Books, 1990) (although she does cover sterilization abuse, 433–36). Charlotte Rutherford, "Reproductive Freedoms and African American Women," *Yale Journal of Law and Feminism* 4 (1992): 273–75, omits the Relfs. A thumbnail sketch appears in Angela Y. Davis, *Women, Race, and Class* (New York: Vintage Books, 1981), 215–16. A more complete treatment appears in Donald T. Critchlow, *Intended Consequences: Birth Control, Abortion, and the Federal Government in America* (New York: Oxford University Press, 1999), 144–47. Jennifer Nelson provides a partial (and error-riven) account in *Women of Color and the Reproductive Rights Movement* (New York: New York University Press, 2003), 65–67. Rebecca Kluchin's excellent *Fit to Be Tied: Sterilization and Reproductive Rights in America, 1950–1980* (New Brunswick, N.J.: Rutgers University Press, 2009) provides the most complete account and advances complementary arguments to this essay.

5. Dean J. Kotlowski, "Deeds versus Words: Richard Nixon and Civil Rights Policy," *New England Journal of History* 56 (Winter 1999–Spring 2000): 122–44; Alan J. Matusow, "Economics, Politics, and the Limits of Presidential Power: The Case of Richard Nixon," ibid., 90–100.

6. Edward J. Larson and Leonard J. Nelson III, "Involuntary Sexual Sterilization of Incompetents in Alabama: Past, Present, and Future," *Alabama Law Review* 43 (1992): 399–444, at 403.

7. Allan Chase, *The Legacy of Malthus: The Social Costs of the New Scientific Racism* (New York: Alfred A. Knopf, 1975), 382–87; Critchlow, *Intended Consequences,* 150–66.

8. "President Signs Birth Curb Bill," *New York Times,* December 27, 1970.

9. See "A bill to provide for the sterilization of females who give birth to certain illegitimate children. . . . ," House Bill 494, February 12, 1960, *Virginia Bills, House* (1960); Gregory Michael Dorr, *Segregation's Science: Eugenics and Society in Virginia* (Charlottesville: University of Virginia Press, 2008), 211–13; and Julius Paul, "The Return of Punitive Sterilization Proposals: Current Attacks on Illegitimacy and the AFDC Program," *Law and Society Review* 3 (August 1968): 77–106.

10. Steven Noll, *Feeble-Minded in Our Midst: Institutions for the Mentally Retarded in the South, 1900–1940* (Chapel Hill: University of North Carolina Press, 1995), 19, 75; James W. Trent, *Inventing the Feeble Mind: A History of Mental Retardation in the United States* (Berkeley: University of California Press, 1994), chapter 5. Noll downplays the salience of race in southern eugenics. I believe Noll overstates his case for the early period; undoubtedly, by the 1960s and 1970s many whites used specious claims of black feeblemindedness to justify the extension of sterilization. See Gregory Michael Dorr, "Defective or Disabled? Race, Medicine, and Eugenics in Progressive Era Virginia and Alabama," *Journal of the Gilded Age and Progressive Era* 5 (October 2006): 377. See also Johanna Schoen, *Choice & Coercion: Birth Control, Sterilization, and Abortion in Public Health and Welfare* (Chapel Hill: University of North Carolina Press, 2005).

11. See generally Premilla Nadasen, *Welfare Warriors: The Welfare Rights Movement in the United States* (New York: Routledge, 2005).

12. Nixon quoted in Kotlowski, "Deeds versus Words," 122–23.

13. John Ehrlichman and Nixon, quoted in Simone Caron, "Richard M. Nixon: The 'Problem of Population' versus 'The Sanctity of Human Life,'" *New England Journal of History* 56 (Winter 1999–Spring 2000): 102–104.

14. Nixon quoted in H. R. Haldeman's diary entry for Monday, April 28, 1969 in Kevin L. Yuill, *Richard Nixon and the Rise of Affirmative Action: The Pursuit of Racial Equality in an Era of Limits* (Oxford: Rowman and Littlefield, 2006), 124.

15. Richard M. Nixon, "Welfare Reform: Shared Responsibility," in *Welfare: A Documentary History of U.S. Policy and Politics*, ed. Gwendolyn Mink and Rickie Solinger (New York: New York University Press, 2003), 314–17.

16. "Nixon: The First Year of His Presidency," *Congressional Quarterly* (1970): 75-A; Jill Quadagno, *The Color of Welfare: How Racism Undermined the War on Poverty* (New York: Oxford University Press, 1994), 118; Michael B. Katz, *In the Shadow of the Poorhouse: A Social History of Welfare in America* (New York: Basic Books, 1996), 277; Nadasen, *Welfare Warriors*, 171–72, 176; Felicia Kornbluh, *The Battle for Welfare Rights: Politics and Poverty in Modern America* (Philadelphia: University of Pennsylvania Press, 2007), 146–47.

17. Nadasen, *Welfare Warriors*, 178–82.

18. David Greenberg, *Nixon's Shadow: The History of an Image* (New York: W. W. Norton, 2003), 304–37.

19. Nixon quoted in Bruce J. Schulman, *The Seventies: The Great Shift in American Culture, Society, and Politics* (New York: Free Press, 2001), 23. My discussion of Nixon's presidency derives from Schulman, Michael Kazin, *The Populist Persuasion* (New York: Basic Books, 1995), chapters 9 and 10, and Thomas Byrne Edsall and Mary D. Edsall, *Chain Reaction: The Impact of Race, Rights, and Taxes on American Politics* (New York: W. W. Norton, 1992).

20. H. R. Haldeman paraphrasing Moynihan, quoted in Kotlowski, "Deeds versus Words," 125.

21. Critchlow, *Intended Consequences*, 89–90; Nadasen, *Welfare Warriors*, 217–18.

22. Critchlow, *Intended Consequences*, 92.

23. Wesley J. Hjornevik, "OEO Instruction—Family Planning Activities," May 18, 1971, Exhibit A; George Contis, M.D., to All OEO Community Action Agency Directors, June 28, 1971, Exhibit B in Relf Files.

24. Simone Caron makes a similar point. Caron, "Richard M. Nixon: The 'Problem of Population,'" 105.

25. "Nominee to O.E.O. Backs Aid to Poor," *New York Times,* July 21, 1973, 25; "Health Official Quits in Protest," *New York Times,* June 29, 1973, 28.

26. "'Cap the Knife' Goes to Work," *New York Times,* July 8, 1973, 4.

27. United States Senate, *Transcript of Proceedings before the Subcommittee on Health of the Committee on Labor and Public Welfare* (Washington, July 10, 1973), Relf Files, 92–107 (quotations 97, 98, 101); "Testimony Conflicts in Sterilization Case," *Washington Post,* July 11, 1973, A2; Leon Cooper, M.D., to Warren Hern, M.D. M.P.H., "Unauthorized External Communications," undated, Exhibit G; and Hern to Cooper, May 8, 1972, Exhibit H, in *Relf v. Weinberger,* Civil Action 1557-73, "Class Action" pleading, Relf Files.

28. "Guidelines Found on Sterilization," *New York Times,* July 7, 1973, 5. For more on Hern, see "The Last Abortion Doctor," *Esquire* (online edition): http://www.esquire .com/features/abortion-doctor-warren-hern-0909.

29. "H.E.W. Chief Issues Guidelines to Protect Rights of Minors and Others in Sterilization Cases," *New York Times,* July 20, 1973, 32; Critchlow, *Intended Consequences,* 145–46.

30. *Roe v. Wade* 410 U.S. 113 (1973), argued December 13, 1971; reargued October 11, 1972; decided January 22, 1973.

31. Charlie Savage, "On Nixon Tapes, Ambivalence over Abortion, Not Watergate," *New York Times,* June 24, 2009.

32. "O.E.O Cuts Off Funds in Sterilization of Girls," *New York Times,* June 29, 1973, 28; Bill Kovach, "H.E.W. Head Curbs Sterilization Aid," *New York Times,* July 6, 1973, 54.

33. Phillips quoted in "Testimony Conflicts in Sterilization Case," *Washington Post,* July 11, 1973, A2.

34. Nicholas von Hoffman, "Life and Liberty in Alabama," *Washington Post,* July 4, 1973, D1 and D4.

35. Philip R. Reilly, *The Surgical Solution: A History of Involuntary Sterilization in the United States* (Baltimore: Johns Hopkins University Press, 1991), 94.

36. Schoen, *Choice & Coercion,* 28–29, 32.

37. Virginia passed the first law making informed, voluntary sterilization explicitly legal in 1962. See Dorr, *Segregation's Science.* See also "Cost of Abortions, Vasectomies and Pills Ruled Tax Deductible," *New York Times,* April 11, 1973, 52. In the next decade, sterilization quietly became the birth control method of choice for one in six American families. See Harold M. Schmeck Jr., "More in U.S. Rely on Sterilization," *New York Times,* March 25, 1974, 17; *Relf v. Weinberger* 372 F. Supp 1196 (1974) at 1199; Jane E. Brody, "Study Finds Sterilization Gains Fastest of Birth-Curb Methods," *New York Times,* May 5, 1976, 22; and Gordon, *Woman's Body,* 432. On Nixon's political backpedaling, see Critchlow, *Intended Consequences,* 166–73.

38. Increased access to reliable contraception often resulted from being labeled irresponsible or unfit. Andrea Tone, *Devices and Desires: A History of Contraceptives in America* (New York: Hill and Wang, 2001), 85–87, 144–45, 268–69; Schoen, *Choice & Coercion,* 32, 41–47.

39. Bond quoted in "Racism, Ethics, and Rights at Issue in Sterilization Case," *New York Times,* July 2, 1973, 10.

40. *Skinner v. Oklahoma,* 316 U.S. 535 (1941), 536, quotation 541. Skinner invalidated Oklahoma's eugenic law mandating sterilization for certain classes of criminals without overturning *Buck.* A constitutional right to privacy was first articulated in *Griswold v. Connecticut,* 379 U.S. 479 (1965), which affirmed married couples' right to

practice contraception, a right extended to the unmarried in *Eisenstadt v. Baird,* 405 U.S. 438 (1972).

41. Mark Haller, *Eugenics: Hereditarian Attitudes in American Thought* (New Brunswick, N.J.: Rutgers University Press, 1963); Kenneth Ludmerer, *Genetics and American Society* (Baltimore: Johns Hopkins University Press, 1972). Daniel J. Kevles posited a subtle transformation of eugenics, from unscientific, racially bigoted "mainline" eugenics to a nonracist "reform" eugenics that focused on population policy in *In the Name of Eugenics* (New York: A. A. Knopf, 1985). Allan Chase first challenged the traditional interpretation in his polemical *Legacy of Malthus.*

42. Schoen, *Choice & Coercion,* 103–11; Diane B. Paul, "The Eugenic Origins of Medical Genetics," in *The Politics of Heredity: Essays on Eugenics, Biomedicine, and the Nature-Nurture Debate* (Albany: State University of New York Press, 1998), 133–56.

43. Edward J. Larson, *Sex, Race, and Science: Eugenics in the Deep South* (Baltimore: Johns Hopkins University Press, 1994), 60–67, 105–107, 139–52.

44. The sterilization clause was buried in a statute establishing a home for the feebleminded. See Julius Paul, ". . . Three Generations of Imbeciles Are Enough . . .: State Eugenic Sterilization Laws in American Thought and Practice," manuscript (Washington, D.C., 1965), 240, available at http://BuckvBell.com.

45. Larson and Nelson, "Involuntary Sexual Sterilization," 414–15.

46. Although most scholars emphasize the disproportionate sterilization of women, Alabama sterilized 129 men and 95 women. Paul, ". . . Three Generations of Imbeciles Are Enough . . . ," 246. On targeting women, see Reilly, *Surgical Solution,* 94–95; Wendy Kline, *Building a Better Race: Gender, Sexuality, and Eugenics from the Turn of the Century to the Baby Boom* (Berkeley: University of California Press, 2001), 53–56.

47. "Report of the Committee on Mental Hygiene," *Transactions of the Medical Association of the State of Alabama,* 1935; Reilly, *Surgical Solution,* 97.

48. James N. Baker, "Eugenics and Human Sterilization," *Journal of the Medical Association of the State of Alabama* (July 1934): 17–18; Larson and Nelson, "Involuntary Sexual Sterilization," 417–18. Baker exemplified many American eugenicists who applauded the Nazis. Stefan Kühl, *The Nazi Connection: Eugenics, American Racism, and German National Socialism* (New York: Oxford University Press, 1994).

49. Alabama House Bill 87, 1935 reg. sess. (introduced June 18, 1935), 92 *Journal of the House of Representatives of Alabama* 143 (1935): 633–36.

50. Partlow quoted in Larson and Baker, "Involuntary Sexual Sterilization," 424. Partlow sterilized all parolees in this period—224 people in 16 years, an average of about 1 per month, hardly "liberal" practices in an institution with hundreds of inmates.

51. The University of Alabama catalog shows eugenics entering Alabama's medical curriculum as early as 1902's Genetic Psychology. By 1915, the biology department's Evolution, Heredity, Genetics offered eugenics. University biologists supported efforts to expand the state's sterilization program in 1935. "Biologist Sees Control of Human Defectives Only Solution of Race," *Montgomery Advertiser,* June 30, 1935, 5. Academic support was not surprising. See Dorr, *Segregation's Science,* chapters 1–3, 6; Steven Selden, *Inheriting Shame: The Story of Eugenics and Racism in America* (New York: Teacher's College Press, 1999).

52. Larson and Nelson, "Involuntary Sexual Sterilization," 425–26.

53. This late date refers to Virginia. In the 1979 case *Hudson v. Hudson,* the Alabama Supreme Court held that courts did not have "'the power to order a surgical sterilization'

even where the operation was in the incompetent's best interest." It reserved the right of courts to order sterilization in the case of medical necessity. Larson and Nelson, "Involuntary Sexual Sterilization," 431–32.

54. See "Aide Says 11 May Have Been Sterilized," *New York Times,* July 3, 1973, 6; Bill Kovach, "H.E.W. Head Curbs Sterilization Aid," *New York Times,* July 6, 1973, 54.

55. Schoen, *Choice & Coercion,* 75–76.

56. See "Inquiry in South Carolina," appended at end of "H.E.W. Chief Issues Guidelines to Protect the Rights of Minors and Others in Sterilization Cases," *New York Times,* July 20, 1973, 32; and "3 Carolina Doctors Are under Inquiry in Sterilization of Welfare Mothers," *New York Times,* July 22, 1973, 30; "Sterilization of Black Mother of 3 Stirs Aiken, S.C.," *New York Times,* August 1, 1973, 27.

57. Elena R. Gutierrez, "Policing 'Pregnant Pilgrims': Welfare, Health Care, and the Control of Mexican-Origin Women's Fertility," in *Women, Health, and Nation: The U.S. and Canada since 1945,* ed. Georgina Feldberg, Molly Ladd-Taylor, Alison Li, and Kathryn McPherson (Ithaca, N.Y.: Cornell University Press, 2003), 384.

58. "Amended Complaint No. 1," Civil Action No. 1557-73 (August 10, 1973), Relf Files; *Walker v. Pierce* 560 F. 2d 609 (1977).

59. Stephen Trombley, *The Right to Reproduce* (London: Weidenfeld and Nicolson, 1988), 177.

60. Rickie Solinger, *Pregnancy and Power: A Short History of Reproductive Politics in America* (New York: New York University Press, 2005), 194–200; Gutierrez, "Policing 'Pregnant Pilgrims,'" 379–403; and Jane Lawrence, "The Indian Health Service and the Sterilization of Native American Women," *American Indian Quarterly* 24 (Summer 2000): 400–419.

61. "Statement of Bernard Lloyd Rosenfeld, M.D., Ph.D.," January 23, 1974, Relf Files.

62. Health Research Group, "A Health Research Group Study on Surgical Sterilization: Present Abuses and Proposed Regulations" (Washington, D.C.: Health Research Group, October 1973), 2, 7, and 9.

63. *Wyatt v. Aderholt* 368 F. Supp. 1382 (M.D. Ala. 1973) and *Wyatt v. Aderholt* 368 F. Supp. 1383 (M.D. Ala. 1974). Inmates had filed suit over inhumane treatment, including allegedly continued nonconsensual sterilization.

64. "Clinic Defends Sterilization of 2 Girls, 12 and 14," *New York Times,* June 28, 1973, 14. These assertions are contradicted by the girls' medical records, which contain both "signed" consent forms from Mrs. Relf authorizing the clinic to provide the girls with the birth control pill and indications that Depo-Provera injections had been administered to each girl at least twice. See "Medical Records," Relf Files.

65. Assistant Secretary of Health Charles C. Edwards, M.D., to The Secretary, DHEW, memorandum, July 3, 1973, Relf Files; "Aide Says 11 May Have Been Sterilized," *New York Times,* July 3, 1973, 6; "H.E.W. Head Curbs Sterilization Aid," *New York Times,* July 6, 1973, 54.

66. Bruce Nichols, "Whites in State Obtained Half of 82 Sterilizations," *Montgomery Advertiser,* June 29, 1973, 9. Later, health department officials acknowledged that this number did not include eight other Alabama agencies, similar to the MCAC, that provided sterilization services. J. Paul Till, "Abortion, Sterilizing Utilized in Alabama," *Montgomery Advertiser,* July 2, 1973, 1. Twenty-six men were sterilized through this program.

67. Bruce Nichols, "Sterilization Case 'Patient's Choice,'" *Montgomery Advertiser,* June 28, 1973, 1, 2.

68. Bruce Nichols, "Staff of Clinic Stunned by Suit," *Montgomery Advertiser,* June 29, 1973.

69. Gregory Michael Dorr, interview with Joseph Levin, August 22, 2007, Southern Poverty Law Center. [Tape in author's possession.] The doctor who sterilized the Relfs had, years earlier, delivered Joseph Levin.

70. Joseph J. Levin Jr. to Steven A. Becker, January 6, 1975, Relf Files.

71. Dorothy I. Height to Julian Bond, July 9, 1973; and Joseph J. Levin Jr. to Dorothy Height, August 3, 1973, Relf Files.

72. See "Racism, Ethics, and Rights at Issue in Sterilization Case," 10. For an extensive discussion of "assimilationist eugenics" among blacks, see Dorr and Logan, chapter 4 of this volume.

73. Paul A. Lombardo, "Three Generations, No Imbeciles: New Light on *Buck v. Bell," New York University Law Review* 60 (April 1985): 51; Michele Mitchell, *Righteous Propagation: African Americans and the Politics of Racial Destiny after Reconstruction* (Chapel Hill: University of North Carolina Press, 230–36; and Gregory Michael Dorr, "Beyond Racial Purity: African Americans and Integrationist Eugenics" (paper presented at 2003 Organization of American Historians conference, April 2003).

74. Nichols, "Staff of Clinic Stunned," 2.

75. M. Cordell Thompson, "Genocide: Black Youngsters Are Sterilized by Alabama Agency," *Jet* (July 19, 1973): 13.

76. Jack Slater, "Sterilization: Newest Threat to the Poor," *Ebony* 28 (October 1973): 156.

77. Susan L. Smith, "Neither Victim nor Villain: Eunice Rivers and Public Health Work," in *Tuskegee's Truths: Rethinking the Tuskegee Syphilis Study,* ed. Susan M. Reverby (Chapel Hill: University of North Carolina Press, 2000), 348–64; Edward H. Beardsley, *A History of Neglect: Health Care for Blacks and Mill Workers in the Twentieth-Century South* (Knoxville: University of Tennessee Press, 1987); Susan L. Smith, *Sick and Tired of Being Sick and Tired: Black Women's Health Activism in America, 1890–1950* (Philadelphia: University of Pennsylvania Press, 1995).

78. Thompson, "Genocide," 12.

79. Mrs. Emily Young, the African American founder of the MCAC's family planning clinic (1967), noted that she had hired Mrs. Dixon as her successor because "the two shared the same philosophy, which emphasizes positive family planning rather than negative birth control." This formulation mirrors the distinction between positive eugenics and negative eugenics, reflecting the discursive shift between older racist and classist eugenics to modern "reform" eugenics. Nichols, "Staff of Clinic Stunned," 2; Kevles, *In the Name of Eugenics,* 88, 173–76.

80. *Wyatt v. Aderholt* 368 F. Supp. 1382 (1973), 1383; *Wyatt v. Aderholt* 368 F. Supp. 1383 (1973), 1384. See Larson and Nelson, "Involuntary Sexual Sterilization," 438–39.

81. Chase, *Legacy of Malthus,* 16.

82. *Relf v. Weinberger* 372 F. Supp. 1196 (1974), 1199.

83. Ibid., 1203.

84. Ibid., 1204.

85. Ibid.

86. Gordon, *Woman's Body,* 435.

87. The 1978 National Organization for Women convention condemned the waiting period as an intolerable infringement on reproductive freedom. Ibid., 434.

88. Melvin Belli to Morris Dees, June 21, 1977, Relf Files.

89. Gregory Michael Dorr, interview with Joseph Levin.

90. Robert E. McGarrah Jr. to Caspar Weinberger, January 20, 1975; Robert E. McGarrah Jr., "Sterilization without Consent: Teaching Hospital Violations of HEW Regulations" (unpublished report, Public Citizen, January 21, 1975), 1.

91. Bogue and Sigelman, "Sterilization Report," *Family Planning Perspectives* 11 (November–December 1979): 366–67.

92. *Walker v. Pierce* 560 F. 2d (1977), 613.

93. "Too Many Babies, Too Many Barriers," *Washington Post,* February 19, 1998, 1, 11.

PART FOUR

Eugenics in the Human Genome Era

NINE

Are We Entering a "Perfect Storm" for a Resurgence of Eugenics? Science, Medicine, and Their Social Context

LINDA L. MCCABE AND EDWARD R. B. MCCABE

The purpose of this volume is to consider the history and legacy of eugenics. No extant scientific or medical community is more conscious of the burden of the eugenic bequest than human and medical geneticists, since our discipline is historically rooted in eugenics and the eugenic movement.[1] Perhaps in response to this disciplinary origin, geneticists often avoid discussions of eugenics, but such avoidance ignores the risk of eugenic resurgence and may in fact foster its reemergence by avoiding notice of the scientific, medical, and social factors that are realigning for the "perfect storm."

FACTORS CONTRIBUTING TO A POSSIBLE RESURGENCE OF EUGENICS

The following proposals appeal to significant portions of the U.S. population in the early twenty-first century. Taxpayers are distressed about spending for the poor, the mentally ill, and the incarcerated. An increased police presence is required to promote the rule of law. The citizenry are concerned about being overwhelmed by recent immigrants and their offspring, and many would support immigration reform that is more selective. But these were early twentieth-century positions espoused by Charles Davenport, who was the founding director of the Cold Spring Harbor Laboratory in 1904 and its Eugenics Record Office in 1910.

Garland Allen argues that many of the economic and social influences that led to the American Eugenics Movement are present today, and these influences exist in a similar scientific context that is filled with genetic determinism. The early decades of the twentieth century, during which eugenics prospered in the United States, were turbulent socially, economically and politically. One response to this turbulence was Progressivism, which utilized a scientific approach to planning and management, engaging experts and managers to address rapid change and improve efficiency, initially in the industrial sector and eventually in government. These features had strong appeal to proponents of eugenics, who argued that science could be harnessed to improve genetic outcomes, and careful management of human breeding would be more efficient for society economically.[2]

The parallels are obvious between Davenport's social and economic positions described above and similar proposals heard today. Allen shows additional resemblances between the early twenty-first and early twentieth centuries and raises serious concerns:

> I would like to suggest that in the United States, immersed as we are in the present economic era of cutbacks and "bottom-line" mentality, we are on the brink of revisiting a mistake of the past, that is, regarding certain people as too expensive to maintain, and using genetic arguments to justify inhumane solutions in the name of efficiency.[3]

Allen adds:

> We seem to be increasingly unwilling to accept what we view as imperfection in ourselves and others. As health care costs skyrocket, we are coming to accept a bottom-line, cost-benefit analysis of human life. This mindset has serious implications for reproductive decisions. If a health maintenance organization (HMO) requires in utero screening, and refuses to cover the birth or care of a purportedly "defective" child, how close is this to eugenics? ... If eugenics means making reproductive decisions primarily on the basis of social cost, then we are well on the road.[4]

The rhetoric of genetic determinism and the perceived power of genomics could fuel the resurgence of eugenics, though undoubtedly with a different name, and therefore it is essential to challenge such deterministic thinking whenever and wherever it arises.

In this era of political and social polarization, reinvention of market economies, and a desire to maximize efficiency in all areas of our lives, will we be seduced into believing that science will be our salvation? Are we entering an era of neo-progressivism that could set the stage for the resurgence of eugenics?

SCIENTIFIC CLAIMS AND EXPECTATIONS

The media, public, and patients have exhibited a remarkable interest in the results of the Human Genome Project.[5] A physician, Sandra Sabatini, asked her patients, "What are your expectations of the Human Genome Project?" and summarized the responses: "Cure diabetes and cancer"; "Prevent renal [kidney] failure, heart disease, and birth defects"; "Avoid hypertension, obesity, and osteoporosis"; and "Prolong life." She added, "Most [of her patient-respondents] believe that these problems are either now solved or near to being solved." How have these expectations been shaped?[6]

Horace Freeland Judson argues that the public's understanding of the Human Genome Project is distorted by the language used to describe it.

> Look at the phrase—or marketing slogan—"the human genome project." In reality, of course we have not just one human genome but billions. At the level of genes, the project promises a useful consensus, but at the level of sequences of nucleotides, variability is great and important.... Then, too, the entire phrase—the human-genome project: singular, definite, with a fixed end-point, completed by 2000, packaged so it could be sold to legislative bodies, to the people, to venture capitalists. But we knew from the start the genome project would never be complete.[7]

He concludes that "genes act in concert with one another—collectively with the environment" and "for ourselves, for the general public, what we require is to get more fully and precisely into the proper language of genetics."[8] A sloppiness of language and miscommunication by the scientists results in an excessively reductionist and determinist perception of the products of genetic and genomic research on the part of the public. The complexity of biology denies simple reductionist explanations for disease based on a specific mutation in a single gene and demands a synthetic view of gene products embedded within biological networks.[9]

THE HUMAN GENOME PROJECT
BECOMES AN ICON

The Human Genome Project would become an icon of the last decade of the twentieth century and the first years of the twenty-first century, and who better to champion this iconic status than James Watson, for he had already accomplished this for the double helical structure of DNA, as chronicled by Soraya de Chadarevian.[10] De Chadarevian noted that the photographs of Watson and Crick with their model of DNA are now inseparable from their 1953 publication in which they deduced the structure of DNA.[11] Those photographs, however, remained in deep obscurity until the publication of Watson's best seller, *The Double Helix*, in 1968.[12] In September 1988, the National Institutes of Health established the Office of Human Genome Research with Watson as its leader. Watson would sell the Human Genome Project to the American people and gain funding from Congress with statements like his 1989 quote in *Time* magazine: "We used to believe our destiny was in the stars; now we know in large measure our fate is in our genes."[13]

Such hyperbole as Watson's would become the public language of the Human Genome Project, sprinkled generously with metaphors and other comparisons.[14] At times these metaphors can take on a life of their own. For example, following the announcement of the "working draft" of the human genome sequence on June 26, 2000, in a joint press conference linking President Bill Clinton in the White House with Prime Minister Tony Blair at 10 Downing Street, the "book of life" metaphor, though not used in Clinton's speech, was found in seven of nine major British newspapers.[15] Symbolic communication in this manner may insert additional meaning with historical references, for example, bringing back "old metaphors of the eugenics movement."[16]

GENETIC DETERMINISM IS ONE MESSAGE OF THE HUMAN GENOME PROJECT

The hyperbolic and metaphoric communications of the Human Genome Project often contain genetic deterministic meanings, implying a reduction of an individual's identity, future, and fate to that person's genes. Consider for a moment the implications of the rhetoric stating that an individual's fate is determined by his/her genomic sequence: since the individual's genome is determined at the time of fertilization and the sequence is set at that moment, then the person's fate would be immutable following that instant. But we know that is not the case, for it is not so much how the sequence is written as how it is read. Environmental influences effect chemical changes in the DNA to silence specific genes so that these genes are not read, and these chemical modifications may be maintained permanently for the life of the individual and perhaps even beyond into the next generation(s). Therefore, an individual's experiences, or perhaps even the experiences of their progenitors, can influence the way in which his/her genome is read throughout his/her life span.

Eugenics utilized genetic determinism to simplify the biological basis of behavior and other exceedingly complex characteristics and generalized to entire ethnocultural groups these "biologically" constructed, perceived deviations from acceptable norms. If one recognizes the complexity of biology and the plasticity of genomic expression, then the simplistic foundations of eugenics must crumble. Lionel Penrose, Galton Professor of Human Genetics at the University College London, described the lack of scientific support for eugenics in his Presidential Address to the Third International Congress of Human Genetics in 1966. He stated, "Eugenics was based upon arbitrary valuations of individuals and social groups, supported by unjustified and premature assumptions about the nature of hereditary influences," and "At the moment . . . our knowledge of human genes and their action is still so slight that it is presumptuous and foolish to lay down the positive principles for human breeding."[17] Penrose made these statements in 1966, but more than 30 years later the Dutch philosopher of the mind, Huib Looren de Jong, said the position "that genes determine development, inevitably leading towards certain types of behavior or personality" is a consequence of "the Gene Myth: the view that our nature (or even our fate) is in the genes, that genes determine behavior like a puppeteer his puppets."[18] Thus the reductionism required for the simplistic interpretation demanded by eugenics is fundamentally flawed because of the complexity of biology and its mutability by its intimate and ongoing interactions with the environment. Genetic determinism drives this simplification, because genetic determinism cannot survive the nuanced reality of biology.

GENETIC DISCRIMINATION

Eugenics, with its arbitrary and fallacious use of an individual's or group's perceived genetic information in decisions regarding social roles (e.g., employment) and biological roles (e.g., reproduction) is one form of genetic discrimination, and both eugenics and genetic discrimination are driven by genetic determinism. This arbitrariness is due in part to the incompleteness of genetic information, the selective use of that information, and the privileging of certain traits with undeserved genetic power. For example, racial and genetic discrimination are often intertwined within eugenics; but endowing racial groups with profound genetic differences denies the fact that race is a social rather than a biological construct, and there is more genetic variation within than between ethnocultural groups.[19]

Just as economic factors were critical to the development of eugenics, so too are economic considerations key influences in genetic discrimination as currently practiced in the United States for insurance and employment. The concerns expressed by employers and insurers in denying benefits relate to the costs of the individual's or group's genetic disorder(s). But as we will see in the next section, some of these decisions, made on the basis of cost containment, productivity, and efficiency, will be considered to meet the criteria to be classified as eugenics.

Do mechanisms exist to prevent and/or redress genetic discrimination? Some individual states have passed laws intended to block or severely limit this form of discrimination, despite "strong opposition and lobbying by the insurance industry and chambers of commerce."[20] At the federal level, several different strategies currently exist to address the problems caused by genetic discrimination. The Equal Employment Opportunity Commission (EEOC), particularly during the term of Commissioner Paul Miller, interpreted the Americans with Disabilities Act of 1990 to prohibit employment discrimination.[21] On February 8, 2000, President Clinton issued Executive Order 13145, "To Prohibit Discrimination in Federal Employment Based on Genetic Information." The Congress took 13 years to pass genetic nondiscrimination legislation for health insurance and employment. However, this legislation only covers those with a positive genetic test who have not yet developed symptoms; those with symptoms, members of the military, and those applying for life, disability, and/or long-term care insurance are not covered.[22]

Thus the federal government and certain state governments recognize the existence of genetic discrimination and have used various strategies to combat it. More sweeping federal nondiscrimination legislation targeting employment and insurance, such as the recently enacted Genetic Information Nondiscrimination Act, will not protect explicitly against eugenics. Such legislation, however, by creating a stronger legal barrier against genetic discrimination, would create an additional defense against eugenic practices in the United States.

GENOMIC MEDICINE AND EUGENICS

The basis for genomic medicine is care that will be predictive, preventive, and personalized.[23] One might expect such an individualized and anticipatory approach for health care to celebrate all persons and to prevent any discriminatory and eugenic misadventures. Consider, however, that genetic testing of individuals and screening of populations is essential to implement genomic medicine. If the results of these analytical methods are used inappropriately, then the testing could lead to genetic discrimination and even to eugenic practices. It is not the power of the genetic and genomic technologies that is the risk but how the output of these technologies—an individual's test result—is used.

A DEFENSE AGAINST EUGENICS:
RESPECT FOR PERSONAL AUTONOMY
AND REPRODUCTIVE FREEDOM

Principles of personal autonomy and reproductive freedom are at the center of geneticists' discussions of whether or not genetics and genetic testing can be considered eugenic. If eugenics involves "a social action program,"[24] then individual choice in the use of clinical genetics services would be considered by some to block the claims that medical genetics is eugenic as long as that choice is freely made and not coerced.[25] Holtzman would consider the following scenario to be an example of eugenics: a U.S. health insurance carrier, as a requirement for insuring the baby, demanded that the mother have prenatal diagnosis and that if the fetus was affected, she would terminate the pregnancy. The company would only insure a "well-born" offspring and would force the mother's compliance, or else have her and not their shareholders accept the economic consequences if she did not comply. As predicted by Allen, this example of a resurgence of eugenics was driven by economic decisions and a "'bottom-line' mentality."[26]

Governments may also be perceived to exert coercive influences on genetic decision making when these entities pay for the services and are expressly interested in balancing costs and benefits toward the benefit side of the ledger.[27] One example, cited by Charles Epstein, is the report of the state-funded prenatal screening program in California, which he said "troubled me quite a bit."[28] The report states: "It is useful to reflect on the missed opportunities for the avoidance of birth defects."[29] In this study, 49 percent of women with a fetus with a chromosomal abnormality [e.g., trisomy 21 or Down syndrome] and 29.3 percent of women with a fetus with a neural tube defect [e.g., spina bifida] did not elect to terminate this pregnancy. While not overtly threatening reproductive freedom, representatives of California's prenatal genetic testing program expounded a perspective, if not a policy, of what would be the "ideal" approach to utilization by individual women. A less value-based evaluation of the results and statement of the conclusions would have been less threatening to the principles of personal autonomy and reproductive freedom and therefore less facilitative for eugenics. Governmental representatives must be more sensitive to the language they use to evaluate policies and processes in genetics.[30]

IS "PROCREATIVE BENEFICENCE" A FUNDAMENTAL OBLIGATION?

Julian Savulescu argued in favor of the principle that he called "procreative beneficence" and defined as follows: "Couples (or single reproducers) should select the child, of the possible children they could have, who is expected to have the best life, or at least as good a life as others, based on the relevant, available information."[31] He maintained that assisted reproductive technologies (ARTs), including in vitro fertilization (IVF) and preimplantation genetic diagnosis (PGD), permit parents to select not only against disease genes but also for favorable genetic traits in their offspring. IVF involves fertilization of eggs retrieved from the biological mother in a culture dish in the laboratory by the addition of sperm from the biological father, after which the embryos are cultured approximately two days and examined through a microscope, and the healthiest appearing are transferred to the uterus of the biological mother or a surrogate. IVF is the first stage of PGD with examination not limited to microscopic observation, but after the embryo grows to approximately the eight-cell stage, the nucleus of one cell is removed and undergoes genetic testing.

At that stage of embryonic development the loss of a single cell does not compromise subsequent development since any cell can give rise to a complete fetus. PGD was developed to permit parents to identify embryos with mutations or chromosomal abnormalities prior to implantation, rather than having to await prenatal diagnosis later in the pregnancy with the option to abort affected fetuses.

Savulescu stated that the principle of procreative beneficence implied that "couples should employ genetic tests for non-disease traits in selecting which child to bring into existence and that we should allow selection of non-disease genes in some cases even if this maintains or increases social inequality," and added, "In the absence of some other reason for action, a person who has good reason to have the best child is morally required to have the best child." Savulescu opened his article with the statement, "Eugenic selection of embryos is now possible," but presumably this relates to the literal meaning of the term as *well born* or in his terminology *best life*. This presumption is based on subsequent statements that procreative beneficence would not be eugenic, since it would involve decision making at the individual or couple level, whereas eugenics engages populations and interferes with reproduction, for example, by eugenic sterilization, "to promote social goods."[32] Subsequently, Savulescu argued emphatically for individual choice in reproductive decision making and respect for the individual's decision even when there might be strong disagreement with doctors or others.[33]

IS PRENATAL GENETIC DIAGNOSIS LEADING TO DESIGNER BABIES?

Let us now look at experience in the United States and United Kingdom with PGD that was carried out, not only for the benefit of the child being born without a genetic disease, but also for advantage to a sibling. We will consider the cases and also the differences in decision-making processes between the two countries.

The first case of this type involved the Nash family in Englewood, Colorado.[34] Molly Nash, daughter of Lisa and Jack, had a genetic disorder, Fanconi anemia, which is associated with birth defects, including an absence of thumbs, and an anemia affecting red cells, white cells, and platelets (the clotting elements), which eventually leads to leukemia.

The parents sought a geneticist to assist them in having a second child with PGD, one who would not have Fanconi anemia and would have a tissue type that would allow the new baby to be a matching donor for a hematopoietic stem cell transplant to Molly. If the Nashes used PGD and selected only for embryos unaffected by Fanconi anemia, then, by chance, the baby might not be a compatible donor, and even with multiple pregnancies, they might not have a donor to save Molly's life before she succumbed to leukemia. Eventually Lisa and Jack found Yury Verlinsky in Chicago, who agreed to work with them, and Verlinsky recovered 30 embryos, 24 of which were unaffected with Fanconi anemia, and five of these were also HLA compatible. In the fourth IVF cycle, the last of the five unaffected, compatible embryos was implanted in Lisa's uterus, resulting in a pregnancy, and Adam was born in 2000.[35] His cord blood, rich in hematopoietic stem cells, was transfused into Molly and resulted in a successful transplant.

For Adam Nash, PGD assured that he would not have the same genetic disease as his older sister; however, this technology also assured that he would have a characteristic that had no benefit to him and would only be beneficial to another individual, Molly. In the U.S. health care system, driven in general by personal autonomy and individual decision making, and where there is a resistance to regulate "the practice of medicine," the entrepreneurial approach of Lisa and Jack Nash could be rewarded if they could find a geneticist who understood their motivation and would be willing to pursue their wishes. Such is not the case in the more highly regulated medical environment in the United Kingdom (UK).

The UK's Human Fertilisation and Embryology Authority (HFEA) requires individual assessment of each PGD application before it can be performed.[36] Shahana and Raj Hashmi, in Leeds, had a son, Zain, with beta-thalassemia, a severe red blood cell anemia, caused by mutations in the beta-globin gene.[37] The Hashmis planned to work with Simon Fishel, in Nottingham, who agreed to collaborate with Verlinsky to provide their PGD, which would be the first of this type planned in the UK: selecting embryos without the genetic disorder that would also be a compatible match for Zain. The HFEA dedicated a working group to the Hashmi case, which was also reviewed by the HFEA's Ethics and Law Committee; the deliberations of these subgroups would then be presented to the full HFEA for deliberation. Debbie Jaggers, who was the HFEA licensing manager, commented during these deliberations, "There's pressure for us not to move to eugenics and a slippery slope and that side of things, and yet there's also the contrary pressure that there's a right to have [treatment], you know, if it's scientifically feasible, we should therefore have the right to have it."[38] The HFEA issued a license to Nottingham to allow PGD as requested by the Hashmis.

Another family received a very different decision from the HFEA. The Whitakers had a child with Diamond-Blackfan anemia (DBA), which in the majority of patients is not an inherited genetic disorder but is a consequence of a spontaneous or sporadic mutation.[39] Therefore, PGD was considered by the HFEA not to have any value in preventing disease in the child who would result from the pregnancy, and its only purpose would be to create a child with the correct HLA type to serve as a cord blood donor, which "was deemed unacceptable and in violation of the strict terms of the HFE Act."[40] The HFEA's denial of a license for PGD to benefit the Whitaker child was controversial, particularly in light of the opposite decision for the Hashmis.

It is valuable to examine these three cases of PGD that were undertaken to create "designer babies" or "savior siblings" and to compare the processes in the United States and the United Kingdom. Sarah Franklin and Celia Roberts considered the practices of the HFEA to have "created a climate of greater openness and exchange among the IVF community, which in turn is seen to exercise a much more effective means of self-regulation than in the more secretive and competitive context of the United States." They concluded, regarding the actions of the HFEA, "In response to some of these cases, such as that of the Hashmis, the existing limits to treatment have been extended, whereas in response to others, such as that of the Whitakers, the limits have been clarified and reinforced."[41]

Ginny Squires of the HFEA stated, "People [in the UK] want this area of medicine regulated, and they see what goes on in America . . . and they want some kind of boundary put on it."[42] Whereas the addition of new technological approaches at the interface of reproduction and genetics in the United States are more likely to be individual decisions that are judged in the media, the UK has established a centralized regulatory approach to assess the comfort level of the public with such innovations. Since the HFEA is a governmental authority, it is setting state-sanctioned policies for the population and, therefore, depending on the policies it establishes, could be judged as exerting eugenic authority. On the other hand, the more individualized initiative and decision making in the United States might seem to have less risk for interpretation as developing population eugenic policies. Let us now consider whether the U.S. system embodies any less risk for support of eugenic policies.

THE "PRACTICE OF MEDICINE"

The "practice of medicine" is a fundamental concept in the U.S. medical system, and the tensions surrounding the regulation or lack of regulation of this practice and where that regulatory line is drawn are well recognized.[43] The unwillingness to restrict medical practice by individual physicians, for example, in the off-label prescribing of medications and devices, is a position promoted by both the American Medical Association and the U.S. Food and Drug Administration (FDA).[44] An example of an off-label use involves an FDA-approved drug prescribed for a different purpose. The Federal Food Drug and Cosmetic Act (FDCA), originally passed in 1938 by the Congress, indicated it had no intention of broadly regulating the practice of medicine. Subsequent court decisions determined that there were no constitutionally based limits on the ability of the FDA to regulate physicians' practices; however, the FDA, "as a matter of policy, has sought to avoid direct regulation of their activities."[45]

One might conclude that the U.S. model of the individualized practice of medicine, which has been referred to as "authoritarian medicine," has the following implications:

> The term authoritarian medicine promotes the power of medical autonomy, the belief that physicians possess such special knowledge that nonprofessionals cannot properly evaluate them, the belief that physicians are conscientious and do not require supervision, the belief that the profession can be trusted to supervise and discipline itself, and the belief that peer review is the highest authority in the profession.[46]

This authoritarian model of medicine, however, is not as independent and unrestricted as it might seem. The regulation of the practice of medicine occurs primarily at the level of the states.[47] States exert this authority through their medical boards, which, in turn, exert their authority through disciplinary actions. State courts also hear medical malpractice lawsuits by citizens. Disciplinary actions include "loss of insurance coverage, negligence or incompetence, malpractice . . . , impairment by substance abuse or mental illness, sexual misconduct with patients, misrepresentation of credentials, and inappropriate prescribing practices."[48] While these offenses all may appear to be serious, constituting professional misconduct, it must also be recognized that the power of such disciplinary authority could begin to circumscribe the boundaries of medical practice.

Therefore, while at the superficial level U.S. medical practice appears to be more independent than, for example, practice in the UK, there are state-based regulatory authorities that establish the rules of medical practice conduct and misconduct. If the norms of medical conduct should be interpreted to shift toward neo-eugenic practices, or at least do not recognize such practices as misconduct, then such authorities could overtly condone or covertly permit a new form of eugenics.

In addition to state-based regulation of medical practice in the United States, there is also the increasing influence of corporate incentives. We have discussed, above, the impact of health insurers' decisions on patients' choices. The rise of the corporate practice of medicine, and its enabling physician-employees, establishes a new dynamic that may influence the practice of medicine and practice policy. With the rise of the employee-physician in the United States, the view of the physician is changing: "The traditional image of the physician in this country seems to center on engagement between an independent medical professional and his or her patient, applying professional judgment in the application of medical science to the particular circumstances presented, with only the patient's interests at play." However, "As employment becomes a more dominant setting for physicians, the dissonance between the image and reality becomes irreparable."[49]

Alan Berkenwald has stated that the "health care revolution" currently underway "is, arguably, less about health and more about money."[50] Christopher Guadagnino, citing the opinions of William Kissick, has elaborated on the financial reasons for the reasons why clinics and hospitals require closer relationships with and employment of physicians: "technology, the need for cost-effectiveness and for hospital formularies," the latter presumably limiting the choice of the individual physician and reducing the costs of therapy.[51] The changes in the organization of medical practice, therefore, are being influenced significantly by what Garland Allen sees as one of the precursors for the resurgence of eugenics, the bottom-line mentality.[52]

The reorganization of medical practice from the independence of the practitioner to the institutional employment of the physician is having other influences and is most definitely constricting the previously more widely distributive decision making and could change the normative standards for the practice of medicine. For example, Guadagnino, citing the opinions of Alan L. Hillman, has indicated "that greater exposure to peer review among employed physicians may lead to increased quality assurance," and interrelationships "between physicians and medical societies could become even more important in the face of increasing physician employment," for example, by assisting physicians to determine "which employee practices should become norms."[53]

Through these analyses of the current state of the practice of medicine, we see the established authority of institutions such as medical societies and state medical boards and the possibly increasing influence of these institutions driven by the need for adjudication of normative practices. Should the reader think that such institutions would never resort to the support of eugenic policies, it is particularly relevant to this volume that Dr. John N. Hurty, an avid supporter of eugenic sterilization, also served as the secretary of the Indiana State Board of Health and firmly believed that every social problem was biologically determined. Hurty was one of the champions of the first eugenic sterilization law.[54] Therefore, we risk a new eugenics through the confluence of the concept that science contains solutions for social problems, the belief in the soundness of genetic determinism, the desire to seek improved efficiencies and decreased costs in the practice of medicine, and the centralization of authority in the determination of practice norms.

COMMERCIALIZATION OF GENETIC TESTING AND REPRODUCTIVE TECHNOLOGIES

Summarizing the arguments of Novas and Nikolas Rose, Franklin and Roberts state, "The gene has become a major component of our ideas of selfhood, identity, responsibility, and even citizenship in the twenty-first century."[55] Rose maintains, based on his own work and that of Foucault and Franklin, that the "vital politics" of the twenty-first century "is concerned with our growing capacities to control, manage, engineer, reshape, and modulate the very vital capacities of human beings as living creatures . . . , a politics of 'life itself.'"[56] If genes are so critical to our self-understanding and our desire to control our future, then genetic testing and reproductive technologies are the critical mediators to achieve this understanding and control. And these mediators have become major business influences. Let us examine the context of the markets in genes, genetic testing, and ARTs, recognizing that human DNA and genes are body parts.

MARKETS FOR ORGANS AND TISSUES

All human parts have a price. Sales of body parts from cadavers have received a tremendous amount of media attention.[57] According to John M. Broder of the *New York Times,*

"It is illegal to sell cadavers or body parts. But it is legal to charge 'reasonable' fees for collecting, shipping, processing, marketing and implanting tissues from corpses. Tissue from one cadaver can be used in 50 to 100 different experiments or procedures, and a typical body can be worth more than $220,000, medical researchers say."[58]

Solid organ transplantation is coordinated in the United States by the United Network for Organ Sharing (UNOS), and therefore data for these types of transplants are readily available. For solid organ transplants, the UNOS homepage lists 96,935 individuals as candidates on their waiting list, and between January and May 2007, 11,746 transplants were performed from 5,869 living and cadaveric donors.[59] The cost of organ or tissue procurement for transplantation is lowest for hematopoietic stem cells at $18,900. For solid organ transplants, the procurement costs range from a low of $41,700–48,500 for a single lung to a high of $199,200–222,300 for a combination of liver, pancreas, and intestine.[60] A significant number of candidates die each year awaiting their transplants, estimated for kidneys as 3,000–5,000 individuals on the candidate list.[61] Therefore, serious consideration is being given to paying donors for organs, particularly kidneys.[62] Because of the lack of availability of an adequate number of donor kidneys, it is estimated that "the voluntary sale of purchased donor kidneys now accounts for thousands of black market transplants,"[63] not only in the United States but also in other countries, including India.[64]

Tissue and cell samples are essential for biomedical research, and the ownership of these samples remains a controversial topic. The Office of Human Research Protection (OHRP) and the Food and Drug Administration (FDA) are two agencies within the U.S. Department of Health and Human Services (DHHS) that provide guidance to Institutional Review Boards (IRBs) at each university or other institution carrying out research on human subjects. Both the OHRP and FDA assert that the informed consent documents, which must be read, understood, and signed by each research subject, should not indicate that the subjects are giving up any property rights to their tissue samples. However, in the three cases (*Moore v. Regents of The University of California, Greenberg et al. v. Miami: Children's Hospital,* and *Washington University v. Catalona*) that have been taken to court by research subjects to attempt to affirm ownership rights over their research specimens, the decisions have all determined that the samples are so valuable to biomedical research that ownership, including the right to sell the samples and access to as well as products from the specimens, belongs to the research institution. Therefore, these court decisions indicate that tissue samples provided for research are commodities that are owned by the research institutions, with the research subjects giving up property rights once they have provided the specimens.[65]

COMMODIFICATION OF GENES
AND GENETIC TESTING

Since cadaveric body parts for studies and organs and tissues for transplantation and research all have been commodified, is this also true for genes and genetic testing? The answer is a resounding yes. The U.S. Patent and Trademark Office (PTO), which derives its authority from the Constitution, has issued approximately 6,000 gene patents, many of which were based solely on sequence information from the Human Genome Project. This means that to use the sequence of a patented gene, the patent holder could demand payment of a licensing fee to anyone desiring to profit from that gene's sequence information. Patent holders may decide that the best way to increase the value of the intellectual property they own is to be extremely restrictive in licensing, perhaps even preventing the use of the sequence information by anyone, hoping that this will add to the perceived value of the patented gene sequence information. The consequence of such a decision to block access to a patented gene would be that individuals desiring testing for mutations in that gene would not be able to have the testing performed and that, even if testing was permitted, the individuals being tested would pay for the right to have their own DNA analyzed for a mutation in his/her gene. It is estimated that each individual's genome contains approximately 20,000–25,000 genes, and therefore if approximately 6,000 patented genes are owned by some other entity, then roughly 25 percent of each person's gene complement is the property not of that person but of someone else.[66]

Let us consider one measure of the value of this gene-based intellectual property, to determine if it supports the view that genes have been commodified. For if there is insufficient value in this commodity, then it would be hard to argue for commodification of human genes. A website that describes itself as "a publicly funded medical genetics information resource developed for physicians, other healthcare providers, and researchers" lists 618 laboratories testing for 1,430 genetic diseases, of which 1,143 are clinical tests and 287 are research-only tests.[67] These are modest numbers and do not give insight into volume or value of genetic testing. Genzyme Corporation, a major supplier of genetic testing, in its 2007 second quarter earnings report stated that its Genetics segment, fueled in part by clinical trials and diagnostic testing, showed a 21 percent increase in revenue to $73.7 million, which compared with $61.0 million during the same period of 2006; this represented 7.9 percent of their $933.4 million in total revenue for the second quarter of 2007, similar to the 7.7 percent of their $793.4 million in total revenue for that period in 2006,[68] and, in 2007 the Genetics sector of their business had revenues three percentage points better than the overall revenue growth of 18 percent.[69] If there were no quarterly financial aberrations, and for the first two quarters of 2006 and 2007 there did not appear to be any significant differences in revenue, then this would amount to an annual revenue in their Genetic business of at least $280 million.

Another company, Myriad Genetics, has attempted to maintain aggressive control of its breast cancer gene patents and has used these patents to develop its predictive medicine product line to utilize genetic testing to identify cancer before it manifests clinically and to personalize medical care.[70] In 2006, Myriad reported testing the 100,000th individual with its breast cancer gene diagnostic analysis and over $100 million in revenue for its predictive medicine line, an increase of 41 percent over the 2005 revenue. These are only two companies with significant revenues deriving from gene patents and diagnostics, and their fiscal performances indicate significant financial value deriving from the commodification of genes and genetic testing.

COMMERCE IN REPRODUCTIVE TECHNOLOGIES

Artificial Reproductive Technology (ART) represents a variety of approaches to aid women who might not be able to become pregnant without assistance. The classic and most common form of ART is in vitro fertilization (IVF), first performed successfully with the birth in 1978 of Louise Brown in the UK, and involving hormonal hyperstimulation of the biological mother's ovaries to release multiple eggs, recovery of those eggs, mixture of the eggs and biological father's sperm in a culture dish to achieve fertilization, culture of embryos in the laboratory for approximately two days, and placement of one or more embryos into the uterus of a woman with appropriate hormonal preparation to proceed with pregnancy. There are many different scenarios in IVF, for example: the biological mother could carry the pregnancy, or a "surrogate mother" could be used; the eggs could be donated or purchased; the sperm could be provided by a male partner, a donation from a male participant, a purchase from a sperm bank, or even a male who has been declared dead and his next of kin requests collection, referred to as posthumous sperm procurement (PHSP); or the pregnancy could be achieved with frozen embryos created previously by any of the above methods, and thereby belonging to the individual or couple, donated by another, or procured from a third party. The sperm can be injected directly into the cytoplasm of the egg, a process referred to as intracytoplasmic sperm injection (ICSI), if fertilization cannot be otherwise achieved.[71]

Congress passed the Fertility Clinic Success Rate and Certification Act (FCSRCA) in 1992, which required the Centers for Disease Control and Prevention (CDC) in Atlanta to gather annual data on ARTs in the United States. The 2004 CDC report indicated that there were 461 clinics performing ART in the United States during that year, representing a steady increase in clinics since the CDC began collecting these data in 1995. Of these clinics, 411 submitted information for 2004, which included 127,977 ART cycles, 99 percent of which were IVF, and 49,458 infants born resulting from ART cycles performed in that year, which the CDC estimated to represent "slightly more than 1 percent of total U.S. births."[72]

Thus the volume of ARTs is impressive, but what about the commercial value? Using April 2004 pharmaceutical costs, the average national cost per cycle of ART was estimated to be $12,400.[73] Correcting for the number of clinics not reporting results (50 of 461 total, or 10.8 percent), the estimated total number of ART cycles in 2004 would be 143,546.[74] Therefore, the total cost for ART cycles in 2004 would be estimated to be $1.78 billion.[75] While this is a large dollar value, one could argue that it is the service cost of preparing the women for ART. Alternatively, this could be judged to be the investment of $1.78 billion in the ART commodity market that will lead to somewhat more than 1 percent of U.S. births. This tension between services and objects as commodities is one that permeates studies of these types of markets.[76]

The objects in the ART commodity market could be considered to be the gametes, i.e., the eggs and sperm. Among the 143,546 estimated total ART cycles in 2004, in 11.8 percent of the cycles, the "eggs or embryos were donated by another woman,"[77] representing an estimated 16,938 cycles with donor eggs or embryos. Rene Almeling reported that in 2002 two "egg agencies," one of which was among the largest in the United States, charged the "recipient clients an agency fee of $3,500 in addition to the donor's fee and her medical and legal expenses," with the donors' fees ranging from $4,000 to 6,000,[78] and medical fees averaging $12,400.[79] Legal fees are not disclosed, but would be in addition to these other fees. If we use a donor's fee in the middle of the range for these two agencies (e.g., $5,000), and assume one donor per ART cycle, then the sum of the agency, donor, and medical fees would be $20,900. With an estimated 16,938 donor cycles per year and a minimum estimated cost of $20,900/ cycle, the estimated annual total cost of "donated" eggs and embryos to recipient clients for ART cycles would be a minimum of $354 million.

The *2004 ART Success Rate* report does not include data regarding whether the sperm were provided by a partner or friend or purchased from a sperm bank. Almeling, however, described two sperm banks: the larger of the two charged $215/vial and distributed approximately 30,000 vials/year, for a total of $6.45 million/year; and the smaller, a nonprofit agency, charged $175/vial and distributed approximately 400 vials/year, for a total of $70,000/year.[80] Therefore, while we do not know the total annual volume and cost of sperm, nevertheless there are costs to some clients for the procurement of male gametes.

Almeling argues, particularly for eggs in the case of postcycle gifts, "the line between gift and sale [is] indistinguishable." She describes the gendered differences in the markets for eggs and sperm and concludes, "The commodification of the human body can be expected to vary based on the sex and gender of that body, as economic valuations intertwine with the cultural norms in specific structural contexts." Almeling includes the following cautionary note:

> In this market [for eggs and sperm], race and ethnicity are biologized, as in references to Asian eggs or Jewish sperm, and it is one of the primary sorting mechanisms in donor catalogs, along with hair and eye color. This routinized reinscription of race at the genetic and cellular level in donation programs, which as medicalized organizations offer a veneer of scientific credibility to such claims, is worrisome given our eugenic history.[81]

Therefore, we would argue that there is commerce in reproductive technologies, including the gametes themselves, which can reach significant volumes and dollars. Furthermore, this commodification of the foundation of pregnancy and eventual personhood could be argued to fit into the bottom-line mentality that Allen feels is leading us toward a resurgence of eugenics, particularly as individuals seek, for example, the best egg donor to provide the characteristics they desire for their "best born" child or the fertility clinic with the best success rate per ART cycle.

With the intimate relationship between our genes and our identities, we would argue that the commodification of genes, genetic testing, and reproductive technologies influence our concepts of individual identity. Placing a value on these objects and services could be perceived as placing a value on selfhood and identity. Given the deterministic nature of the ART industry, for example, "If you obtain an egg from a woman with an advanced degree, then you will have a brighter child,"[82] and the fact that the ova from the individual with these desired characteristics will be priced higher than others, the marketplace is differentially valuing individuals within society. Moving from the market value of individuals to their social value would seem to be even easier than the reverse: if a person has a defined monetary worth that is less than another, then some would argue (all too easily) that the person belongs to a different and lower social stratum and he/she would be valued less. These factors as well as the racialization of the ART market[83] and the close relationships of economic/social class and racial categorizations with eugenics in the United States are setting the social stage for a return to eugenics.

Additional forces are present that risk placing us even more directly in the path of this perfect storm that could result in a resurgence of eugenics. Genetic determinism, fueled by the Human Genome Project and fundamental to genetic discrimination and eugenics, reduces the individual to his/her genes and oversimplifies the science, ignoring the need for a more synthetic, nuanced, and realistic view of our biological selves. Overselling the power of science to rationalize and to solve social problems was one of the fundamental tenets of Progressivism, and we would argue that the flawed genetic determinism of the "Genomic Era" could lead us toward a neo-progressivism.

If anyone should think that this return to the Progressive Era is pure academic speculation, please consider the following quote from Hillary Rodham Clinton during the July 23, 2007, CNN/YouTube Democratic presidential debate in response to a question about how she would describe her political philosophy:

> I prefer the word "progressive," which has a real American meaning, going back to the progressive era at the beginning of the twentieth century. I consider myself a modern progressive, someone who believes strongly in individual rights and freedoms, who believes that we are better as a society when we're working together and when we find ways to help those who may not have all the advantages in life get the tools they need to lead a more productive life for themselves and their family. So I consider myself a proud modern American progressive, and I think that's the kind of philosophy and practice that we need to bring back to American politics.[84]

It is concerning that Clinton does not know or consider the eugenic baggage carried by the Progressive movement in the early twentieth century, and that the freedoms and rights of individuals she attributes to this movement were limited only to selected "well-born" persons.

How can we prevent being engulfed by this "perfect storm" for the resurgence of eugenics? We would propose that the following efforts could bring relief and move us out of the path of this storm. We must recognize the illegitimacy of genetic discrimination and support genetic nondiscrimination legislation; respect personal autonomy and reproductive freedom; recognize and resist the corrosive nature of commodification of body parts, including genes and gametes, on social structure; guard against policies from organized medicine, governments, and corporations that could enshrine eugenic practices; and educate opinion leaders and the populace regarding the risks of overpromising genetics/genomics and the consequences of determinism and eugenics.

NOTES

1. See generally Daniel Kevles, *In the Name of Eugenics: Genetics and the Uses of Human Heredity* (New York: Harvard University Press, 1985); Kenneth L. Garver and Bettylee Garver, "Eugenics: Past, Present, and Future," *American Journal of Human Genetics* 49 (1991): 1109–18; Allen Buchanan, Dan W. Brock, Norman Daniels, and Daniel Wikler, *From Chance to Choice: Genetics and Justice* (Cambridge: Cambridge University Press, 2000); Nikolas Rose, *The Politics of Life Itself: BioMedicine, Power, and Subjectivity in the Twenty-first Century* (Princeton, N.J.: Princeton University Press, 2007). Many of the issues we analyze in this essay were raised in our earlier book: Linda L. McCabe and Edward R. B. McCabe, *DNA: Promise and Peril* (Berkeley: University of California Press, 2008).

2. Garland E. Allen, "The Social and Economic Origins of Genetic Determinism: A Case History of the American Eugenics Movement, 1900–1940, and Its Lessons for Today," *Genetica* 99 (1997): 77–88, at 78–80; Garland E. Allen, "Is a New Eugenics Afoot?" *Science* 294 (2001): 59–61.

3. Allen, " Social and Economic Origins," 85.

4. Allen, "Is a New Eugenics Afoot?" 61.

5. Dorothy Nelkin, "Molecular Metaphors: The Gene in Popular Discourse," *Nature Review Genetics* 2 (2001): 555–59, at 557; Horace F. Judson, "Talking about the Genome: Biologists Must Take Responsibility for the Correct Use of Language in Genetics," *Nature* 409 (2001): 769; and Tania Bubela, "Science Communication in Transition: Genomics Hype, Public Engagement, Education, and Commercialization Pressures," *Clinical Genetics* 70 (2006): 445–50, at 446.

6. Sandra Sabatini, "Mapping of the Human Genome: What Does It Mean for Our Patients?" *American Journal of the Medical Sciences* 322 (2001): 175–78, at 175.

7. Judson, "Talking about the Genome," 769.

8. Ibid.

9. Charles R. Scriver and Paula J. Waters, "Monogenic Traits Are Not Simple: Lessons from Phenylketonuria," *Trends in Genetics* 15 (1999): 267–72, at 267; Katrina M. Dipple and Edward R. B. McCabe, "Phenotypes of Patients with 'Simple' Mendelian Disorders Are Complex Traits: Thresholds, Modifiers, and Systems Dynamics," *American Journal of Human Genetics* 66 (2000): 1729–35, at 1729; Katrina M. Dipple and Edward R. B. McCabe, "Modifier Genes Convert 'Simple' Mendelian Disorders to Complex Traits," *Molecular Genetics and Metabolism* 71 (2000): 43–50, at 47; Katrina M. Dipple and Edward R. B. McCabe, "Consequences of Complexity within Biological Networks: Robustness and Health, or Vulnerability and Disease," *Molecular Genetics and Metabolism* 74 (2001): 45–50, at 49–50.

10. Soraya de Chadarevian, "The Making of an Icon," *Science* 300 (2003): 255–57.

11. James D. Watson and Francis H. C. Crick, "Molecular Structure of Nucleic Acids," *Nature* 171 (1953): 737–38.

12. James D. Watson, *The Double Helix: A Personal Account of the Discovery of Structure of DNA* (New York: Readers Union, 1968).

13. Patricia Wald, "Future Perfect: Grammar, Genes, and Geography," *New Literary History* 31 (2000): 681–708, attributed to Leon Jaroff, "The Gene Hunt," *Time*, March 10, 1989, 67.

14. Nelkin, "Molecular Metaphors," 558–59.

15. Brigitte Nerlich, Robert Dingwall, and David D. Clarke, "The Book of Life: How the Completion of the Human Genome Project Was Revealed to the Public," *Health: An Interdisciplinary Journal for the Social Study of Health, Illness, and Medicine* 6 (2002): 445–69, at 450.

16. Nelkin, "Molecular Metaphors," 557.

17. Lionel S. Penrose, "The Influence of the English Tradition in Human Genetics," in *Proceedings of the Third International Congress of Human Genetics*, ed. James F. Crow and James V. Neel (Baltimore, 1967): 13–25, at 22–23.

18. Huib Looren de Jong, "Genetic Determinism: How Not to Interpret Behavioral Genetics," *Theory and Psychology* 10 (2000): 615–37, at 621.

19. See generally Noel Ignatiev, *How the Irish Became White* (New York: Routledge, 1995); Neil Foley, *The White Scourge: Mexicans, Blacks, and Poor Whites: Texas Cotton Culture* (Berkeley: University of California Press, 1997); Dorothy Roberts, *Killing the Black Body: Race, Reproduction, and the Meaning of Liberty* (New York: Vintage Books, 1997).

20. Paul R. Billings, "Genetic Nondiscrimination," *Nature Genetics* 37 (2005): 559–60.

21. McCabe and McCabe, *DNA: Promise and Peril*, 185–88.

22. Kathy L. Hudson, M. K. Holohan, and Francis S. Collins, "Keeping Pace with the Times: The Genetic Information Nondiscrimination Act of 2008," *New England Journal of Medicine* 358 (2008): 2661–63; Russell Korobkin and Rahul Rajkumar, "The Genetic Information Nondiscrimination Act: A Half-Step toward Risk Sharing," *New England Journal of Medicine* 359 (2008): 335–37; Mark A. Rothstein, "Putting the Genetic Discrimination Act in Context," *Genetics in Medicine* 10 (2008): 655–56.

23. Linda L. McCabe and Edward R. B. McCabe, "Genetic Screening: Carriers and Affected Individuals," *Annual Review of Genomics and Human Genetics* 5 (2004): 57–69, at 57–58, 65.

24. Allen, "Social and Economic Origins," 79.

25. Neil Holtzman, "Eugenics and Genetic Testing," *Science in Context* 11 (1998): 397–417, at 408–409; Charles J. Epstein, "Is Modern Genetics the New Eugenics?" *Genetics in Medicine* 5 (2003): 469–75, at 474.

26. Allen, "Social and Economic Origins," 85.

27. Allen, "Is a New Eugenics Afoot?" 59–61.

28. Epstein, "Is Modern Genetics the New Eugenics?" 473.

29. George C. Cunningham and D. Gwynne Tompkinson, "Cost and Effectiveness of the California Triple Marker Prenatal Screening Program," *Genetics in Medicine* 1 (1999): 199–206, at 199.

30. Judson, "Talking about the Genome," 769.

31. Julian Savulescu, "Procreative Beneficence: Why We Should Select the Best Children," *Bioethics* 15 (2001): 413–26, at 415.

32. Ibid., 424.

33. Julian Savulescu, "Deaf Lesbians, 'Designer Disability,' and the Future of Medicine," *British Medical Journal* 325 (2002): 771–73.

34. Details of the Nash case may be found in Sarah Franklin and Celia Roberts, *Born and Made: An Ethnography of Preimplantation Genetic Diagnosis* (Princeton, N.J.: Princeton University Press, 2006), 35, 62.

35. T. Scheck, "A Question of Life," Minneapolis Public Radio, October 18, 2000 (http://news.minnesota.publicradio.org/features/200010/18_scheckt_babies/).

36. Accounts of the Hashmi and Whitaker family cases are drawn from Franklin and Roberts, *Born and Made,* 62, 65–66.

37. Technologies used to screen for neonatal blood conditions such as beta-thalassemia are described in Urvashi Bhardwaj, Yao-Hua Zhang, and Edward R. B. McCabe, "Neonatal Hemoglobinopathy Screening: Molecular Genetic Technologies," *Molecular Genetics and Metabolism* 80 (2003): 129–37.

38. Franklin and Roberts, *Born and Made,* 62–63.

39. *Online Mendelian Inheritance in Man* (OMIM), "#105650, Diamond-Blackfan Anemia; DBA."

40. Franklin and Roberts, *Born and Made,* 65–66.

41. Ibid., 65–67.

42. Ibid., 72–73.

43. Barbara J. Evans, "Distinguishing Product and Practice Regulation in Personalized Medicine," *Clinical Pharmacology & Therapeutics* 81 (2007) 288–93, at 288.

44. Nicole Ansani, Carl Sirio, Thomas Smitherman, Bethany Fedutes-Henderson, Susan Skledar, Robert J. Weber, Nathalie Zgheib, and Robert Branch, "Designing a Strategy to Promote Safe, Innovative Off-Label Use of Medications," *American Journal of Medical Quality* 21 (2006): 255–61, at 255.

45. Barbara J. Evans and David A. Flockhart, "The Unfinished Business of U.S. Drug Safety Regulation," *Food and Drug Law Journal* 61 (2006): 753–94, at 775.

46. Alan D. Berkenwald, "In the Name of Medicine," *Annals of Internal Medicine* 128 (1998): 246–50, at 248–49.

47. L. E. Smith, "The Practice of Medicine and Its Interface with Medical Regulation," *Archives of Facial and Plastic Surgery* 1 (1999): 58–59; James N. Thompson, "The Future of Medical Licensure in the United States," *Academic Medicine* 81 (2006): S36–S39, at S38; J. T. Rannazzisi, "The DEA's Balancing Act to Ensure Public Health and Safety," *Clinical Pharmacology and Therapeutics* 81 (2007): 805–806.

48. James Morrison and Peter Wickersham, "Physicians Disciplined by a State Medical Board," *JAMA* 279 (1998): 1889–93, at 1889.

49. Christopher Guadagnino, "The Rise of the Physician Employee," *Physician's News Digest* (1997): 1–7, at 1.

50. Berkenwald, "In the Name of Medicine," 246.

51. Guadagnino, "The Rise of the Physician Employee," 5.

52. Allen, "Social and Economic Origins," 85.

53. Guadagnino, "The Rise of the Physician Employee," 7.

54. Kevin B. O'Reilly, "Confronting Eugenics: Does the Now Discredited Practice Have Relevance to Today's Technology?" *AMNews*, July 9, 2007, (http://www.ama.assn .org/amednews/2007/07/09/prsa0709.htm).

55. Franklin and Roberts, *Born and Made*, 15.

56. Nikolas Rose, *The Politics of Life Itself*, 3.

57. Andrew Murr, "The Body Parts Bandit?" *Newsweek National News* Web Exclusive, May 9, 2007 (http://www.msnbc.msn.com/id/17542038/site/newsweek/).

58. John M. Broder, "UCLA Halts Donations of Cadavers for Research," *New York Times*, March 10, 2004.

59. UNOS (http://www.unos.org/).

60. N. J. Ortner, "U.S. Organ and Tissue Transplant Cost Estimates and Discussion," *Milliman Research Reports* 6 (2005) (www.transplantliving.org/Content Documents/2005Milliman_Report.pdf).

61. Eli A. Friedman and Alan H. Friedman, "Payment for Donor Kidneys: Pros and Cons," *Kidney International* 69 (2006): 960–62.

62. Troyan Brennan, "Markets in Health Care: The Case of Renal Transplantation," *Journal of Law, Medicine, and Ethics* 35 (2007): 249–55, at 249.

63. Eli A. Friedman and Alan H. Friedman, "Payment for Donor Kidneys, 960–62.

64. Sanjay Kumar, "Police Uncover Large Scale Organ Trafficking in Punjab," *British Medical Journal* 326 (2003): 180.

65. McCabe and McCabe, *DNA: Promise and Peril*, 152–58.

66. Ibid., 163–67.

67. GeneTests data as of August 23, 2007 (http://www.genetests.org/).

68. See http://www.genzyme.com/corp/investors/GENZ%20PR-072507.asp.

69. See http://www.genzyme.com/components/highlights/genz_p1_q2_2007 .pdf.

70. See http://www.myriad.com/about/.

71. McCabe and McCabe, *DNA: Promise and Peril*, 195–97.

72. Centers for Disease Control and Prevention, American Society for Reproductive Medicine, and Society for Assisted Reproductive Technology, "2004 Assisted Reproduction Technology Success Rates: National Summary and Fertility Clinic Reports" (Atlanta, 2006): http://www.cdc.gov/art/art2004/508PDF/2004ART_Intro-National-Sum_t508.pdf.

73. Robert L. Gustofson, James H. Segars, and Frederick W. Larsen, "Cost Analysis of Ganirelix versus Luteal-Phase Lupron Downregulation per Assisted Reproductive Technology (ART) Cycle and Pregnancy Achieved," *Fertility and Sterility* 82 (2004): S118–S119, at S118.

74. CDC et al., "2004 Assisted Reproduction Technology Success Rates."

75. Ibid.

76. Rene Almeling, "Selling Genes, Selling Genders: Egg Donors, Sperm Banks, and Medical Marketing Genetic Material," *American Sociological Review* 72 (2007): 319–40, at 321–22.

77. CDC et al., "2004 Assisted Reproduction Technology Success Rates."

78. Almeling, "Selling Genes, Selling Genders," 332;

79. Gustofson et al., "Cost Analysis of Ganirelix," S118.

80. Almeling, "Selling Genes, Selling Genders," 325–34.

81. Ibid., 337.

82. McCabe and McCabe, *DNA: Promise and Peril*, 206.

83. Almeling, "Selling Genes, Selling Genders," 327.

84. CNN 2007 (http://www.cnn.com/2007/politics/07/23/debate.transcript/index.html).

TEN

Modern Eugenics
and the Law

MAXWELL J. MEHLMAN

One hundred years after the enactment of chapter 215 of the Indiana Acts of 1907, the world's first eugenic sterilization law, eugenics might be expected to be a thing of the past.[1] Yet practices that might be considered eugenic persist, and there is good reason to expect them to flourish in the near future. This essay begins by discussing what is meant by "eugenic." It then describes modern practices that might be considered eugenic. It concludes by foreshadowing future developments in technology and public policy that might achieve eugenic objectives.

WHAT COUNTS AS A EUGENIC PRACTICE?

Francis Galton, who originated the term *eugenics* in 1883, defined it as "the science of improving stock, which is by no means confined to questions of judicious mating, but which, especially in the case of man, takes cognisance of all influences that tend in however remote a degree to give to the more suitable races or strains of blood a better chance of prevailing speedily over the less suitable than they otherwise would have had."[2] In 1904, he restated his understanding of eugenics as "the science which deals with all influences that improve the inborn qualities of a race; also with those that develop them to the utmost advantage."[3]

There is widespread agreement now among geneticists that there is no genetic basis for the concept of "race." Members of the same "race" can differ markedly, and people of different races can share more genes in common than people of the same race.[4] Hence we might be tempted to think that, since eugenics is an effort to improve "races" or "strains of blood," there can be no true modern eugenic practices.

219

Moreover, one of the hallmarks of the eugenics movement in the early twentieth century was its belief that many social ills could be attributed to bad genes. As Elof Carlson noted in the first chapter of this volume, the view that inherited physiological imperfections led to social degeneracy was a prerequisite to the enactment of initial eugenic sterilization laws. Toward the end of the twentieth century, however, our understanding of genetic science has become more sophisticated, and there may be less inclination to believe in genetic determinism.

But the conception of eugenics reflected in these positions is too narrow. It does not matter whether genes or social conditions make a person a thief or a prostitute rather than an upstanding citizen. The point is that thieves and prostitutes are thought to be likely to have children who become thieves and prostitutes, while the children of upstanding citizens are more likely to become upstanding citizens. To improve society, the objective then is to limit the former's ability to reproduce relative to the latter. As Dorothy Roberts states in connection with efforts to criminalize drug-addicted women who bear children, "These reproductive punishments are not strictly eugenic because they are not technically based on the belief that crime is inherited; that is, their goal is not to prevent the passing down of crime-marked genes. They are based, however, on the same premise underlying the eugenic sterilization laws—that certain groups in our society do not deserve to procreate."[5]

Furthermore, efforts at genetic improvement need not be aimed at "races" in order to raise many of the same concerns as classic eugenics. As Roberts implies, the point is whether one believes that only certain types of persons should procreate.

Understood in this way, many modern practices can be considered eugenic, or to use Sonia Suter's term, "neoeugenic." As she explains, "although neoeugenics is not identical to the eugenics of yesteryear, many of the same impulses and drives exist today; most notably, the desire to improve the human species and our children through reproductive choices."[6]

Eugenic practices can be either "positive" or "negative." Daniel Kevles states that positive eugenics aims "to foster greater representation in a society of people whom eugenicists [consider] socially valuable." Negative eugenics, on the other hand, seeks "to encourage the socially unworthy to breed less or, better yet, not at all." [7] As Carlson observes, it aims at "reproductive isolation of those considered unfit to reproduce."

Commentators often distinguish between "state-sponsored" and "private" reproductive decisions, suggesting that only the former can be considered to be eugenics. Duster states, for example: "It is imperative to distinguish between state-sanctioned eugenics programs on the one hand, and private, individualized, *personal* decisions that are socially patterned on the other."[8] But the exercise of state power and the dividing line between state-sponsored and private decision making are less clear-cut than Duster implies. At one extreme, the state can prohibit a eugenic practice. In *Skinner v. Oklahoma,* for example, the Supreme Court held that it was unconstitutional for a state to require certain habitual criminals to be sterilized but not others.[9] Another case attacking sterilization was *Relf v. Weinberger,*[10] discussed earlier in this volume by Gregory Dorr. At issue were regulations proposed in 1974 by the Department of Health, Education, and Welfare (now the Department of Health and Human Services) that would have authorized federal family planning funds to pay for the involuntary sterilization of minors and persons with mental deficiencies, and the supposedly voluntary sterilization of consenting adults. The decision was written by Judge Gerhard Gesell of Watergate fame, who ruled that Congress had not authorized the sterilization of incompetent persons and that the regulations did not incorporate safeguards necessary to fulfill a congressional prohibition against the use of family planning funds to coerce indigent patients into agreeing to be sterilized.

At the other extreme, the state may compel a eugenic practice, such as Indiana did when it enacted its 1907 involuntary sterilization law. Between these two extremes, however, there are a number of gradations. For example, the government can provide financial inducements for eugenic practices, such as tax breaks and welfare penalties for having desirable or undesirable children, or damages awarded by the judicial system to families whose physicians failed to prevent the birth of children with disabilities. Even when the government merely refrains from limiting or prohibiting eugenic practices, it might be said to be sanctioning them implicitly. Duster might regard reproductive decisions by parents as "personal," but the government arguably collaborates when it turns a blind eye.

In short, when the law influences reproductive decision making in such a way as to encourage or discourage the birth of specific types of individuals, this is at least suggestive of a eugenic objective. The more overt this objective, or the greater the impact of the law on individual reproductive decisions, the more frankly eugenic the law can be considered. This is not meant to imply that forced sterilization is on par with government inaction in the face of private behavior. Yet as the other chapters in this book demonstrate, society has employed indirect as well as direct means to control reproductive behavior in order to achieve eugenic goals. The memory of the victims of the Indiana law and its successors would be dishonored if we examined the current state of affairs using too narrow a view of this history.

LEGALLY TOLERATED PRACTICES

A number of practices typically are thought of as private because they result from private rather than public decision making. But the fact that the law permits them to take place indicates a measure of public acquiescence, if not approval.

This category includes attempts to achieve what might be considered positive neoeugenic objectives. One example is selective breeding to produce a "better" genetic lineage. This has long taken place with animals, but it also takes place in humans. The most notorious recent case is the Chinese basketball player Yao Ming, whose parents were mated by the Chinese government because of their height and athletic prowess and who was forced to play basketball at an early age.[11] Private forms of selective breeding are rampant in the United States, including mating within certain social circles, such as "coming out" at debutante balls, arranged marriages and semi-arranged marriages in which the couple is brought together by the parents, and most recently, computerized dating services in which participants select one another according to desirable traits. One such service, eHarmony.com, provides information about 29 personal characteristics, including appearance, intellect, industriousness, ambition, family background, education, and character. Those who avail themselves of these dating services can compress investigations that used to take a number of "dates" into the click of a button.

Some of the most glaring selective breeding practices are associated with gamete donation. The Genetics & IVF Institute, for example, provides the following information about egg donors: adult photos, childhood photos, audio interviews, blood type, ethnic background of donor's mother and father, height, weight, whether pregnancies have been achieved, body build, eye color, hair color and texture, years of education and major areas of study, occupation, Scholastic Aptitude Test (SAT) scores and grade point averages, special interests, family medical history, essays by donors, and personality typing based on the Keirsey test, which uses a Jungian approach to classify temperaments into 12 categories, such as "rational and reserved" or "artisan and introspective." The California Cryobank, Inc., provides purchasers of donor sperm with a 26-page donor profile. A company called Fertility Alternatives pays a premium to "exceptional" egg donors. To qualify, the donor must have graduated from a major university, or be currently attending one, preferably Ivy League; have a GPA of 3.0 +; SAT scores of 1350+ or ACT scores of 30+; and have a documented high IQ.

The degree to which the law accommodates these assisted reproductive practices is striking.[12] Although health care is one of the most heavily regulated industries in the country, the only federal law specifically governing in vitro fertilization (IVF) services is the Fertility Clinic Success Rate and Certification Act of 1992, which merely requires infertility clinics to report success rates in a standardized fashion.[13] State regulation is minimal.[14] The criticisms leveled at the recent IVF-assisted birth of octuplets to an unmarried, unemployed woman illustrate this regulatory lacuna.

Conceivably, state law might encourage selective breeding by permitting parents to sue infertility clinics for negligence in failing properly to screen gamete donors. The leading case is *Harnicher v. University of Utah Medical Center*.[15] The Harnichers purchased sperm from a fertility clinic after closely matching the donor to the husband's physical characteristics and blood type and mixing the donor sperm with the husband's, so that the couple could more easily represent any child genetically as their own. The parents sued after learning that the clinic had accidentally substituted sperm from a different donor, resulting in triplets whose eye color, hair color, and blood type did not match the husband's. The court refused to allow the parents to recover damages from the clinic on the grounds that they had not suffered any physical injury and that they could not recover for the emotional distress from "exposure to the truth" about their children's father.[16] The court also rejected the parents' claim that they had been damaged by the fact that children from the donor they actually had selected would have been better looking.[17] So far it would appear that the law was not lending itself to supporting the practice of selective breeding through gamete donation. But the Supreme Court of Utah concluded its opinion with a startling caveat: "The Harnichers do not allege that the triplets are unhealthy, deformed, or deficient in any way. Nor do they claim any racial or ethnic mismatch between the triplets and their parents."[18] If any of these had been the case, presumably the law would have afforded the parents the relief that they were seeking.

While the parents in the *Harnicher* case used selective breeding in the hopes of achieving a positive neoeugenic goal, the law also countenances a number of practices that might be deemed negative neoeugenics. One obvious example is fetal testing followed by aborting the fetus for reasons that do not relate to the mother's health. Another is preimplantation genetic diagnosis (PGD) of fertilized embryos in the course of IVF, where only those embryos with the best genetic endowment are implanted in the uterus and allowed to gestate. The law leaves IVF services essentially unregulated. A third technique is community-based genetic testing for matchmaking purposes, such as screening programs for Tay-Sachs and other recessive genetic diseases prevalent in Jews of Ashkenazi descent.[19] The pioneer program is Chevra Dor Yeshurim (Association of the Upright Generation), which screens Orthodox Jewish adolescents in New York and does not allow a matchmaker to arrange a marriage between two individuals who both carry a recessive mutation.[20]

Finally, there are private family planning initiatives. One is Project Prevention (formerly CRACK). This was started in California in 1997 by a woman who had adopted four children and who wanted to punish their drug-addicted mother. When she failed to persuade the state legislature to criminalize drug-addicted mothers, she created CRACK, which gives drug addicts $200 if they undergo sterilization or use long-term birth control. In 2003, the organization claimed to have 23 chapters nationwide and had paid 907 women, including 361 who had tubal ligations. African Americans and Hispanics accounted for 401 of the participants.[21] At present, there is no state or federal law that would interfere with this project.

LEGALLY FACILITATED PRACTICES

In some instances, the law goes farther than turning a blind eye; it actually aids the practice of neoeugenics. One example is the ability of parents to bring so-called wrongful birth actions. In most jurisdictions, the common law permits parents to bring an action when a child who is born with a defect or disability would not have been born at all but for the negligence of a health care provider. One example is *Molloy v. Meier*, a 2004 case in which the Minnesota Supreme Court held that physicians who negligently failed to diagnose a genetic disorder known as Fragile X in one child could be liable to the parents who conceived another child with the same disorder.[22] The physicians argued that their duty only extended to the first child, whom they had misdiagnosed. The court disagreed, stating:

> In this case, the patient suffered from a serious disorder that had a high probability of being genetically transmitted and for which a reliable and accepted test was widely available. The appellants should have foreseen that parents of childbearing years might conceive another child in the absence of knowledge of the genetic disorder. The appellants owed a duty of care regarding genetic testing and diagnosis, and the resulting medical advice, not only to S.F. [the first child] but also to her parents.[23]

The significance of this type of lawsuit for neoeugenics is obvious. Indeed, the judge in one Michigan case, *Taylor v. Kurapati*, refused to allow the parents to recover damages in a similar case because of the eugenic implications.[24] In *Kurapati*, a child was born with femur-fibula-ulna syndrome, a condition which left her with a missing right shoulder, fusion of the left elbow, missing digits on the left hand, missing femur on the left leg, and short femur on the right.[25] The parents had obtained two ultrasounds during the pregnancy, but neither radiologist had discovered the abnormalities. The parents contended that this was malpractice and that it deprived them of the opportunity to abort the fetus.

In dismissing the complaint, Judge Whitbeck began by explaining that, in order to recover, the parents would have to prove that the costs of raising the child outweighed the benefits. This required the jury to predict, and discount, the child's future potential. As the judge observed, "How, for example, would a hypothetical Grecian jury, operating under Michigan jurisprudence, measure the benefits to the parents of the whole life of Homer, the blind singer of songs who created the Iliad and the Odyssey?"[26]

In addition, the judge felt that permitting recovery in these cases "can slide ever so quickly into applied eugenics."[27] As he explained:

The very phrase "wrongful birth" suggests that the birth of the disabled child was wrong and should have been prevented. If one accepts the premise that the birth of one "defective" child should have been prevented, then it is but a short step to accepting the premise that the births of classes of "defective" children should be similarly prevented, not just for the benefit of the parents but also for the benefit of society as a whole through the protection of the "public welfare." This is the operating principle of eugenics.[28]

After describing the history of eugenics and its culmination in the Holocaust, Judge Whitbeck concluded:

To our ears, at the close of the twentieth century, this talk of the "unfit" and of "defectives" has a decidedly jarring ring; we are, after all, above such lethal nonsense. But are we? We know now that we all have at least five recessive genes but, according to Bowman [James E. Bowman, "The Road to Eugenics," 3 *U. Chic. L. Sch. Roundtable* 491 (1996)], when scientists map the human genome, they will unveil many more potentially harmful genes in each of us. Bowman states that "psychoses, hypertension, diabetes, early- and late-appearing cancers, degenerative disorders, susceptibility genes for communicable diseases, genes for various mental deficiencies, aging genes, and other variations and disorders will be ascertained." Will we then see the tort of wrongful birth extended to physicians who neglect or misinterpret genetic evidence and thereby fail to extend the option of a eugenic abortion to the unsuspecting parents of a genetically "unfit" and "defective" child? Our current acceptance of the wrongful birth tort would require the answer to this question in Michigan to be: yes.[29]

In 2003, the Kentucky Supreme Court reached a similar result in *Grubbs v. Barbourville Family Health Ctr., P.S.C.*[30] The decision involved two cases. In both, parents were told that the results of ultrasounds were normal. In one case, the child was born with spina bifida and hydrocephalus. The child in the other case had a cyst that occupied most of his cranium. As a consequence, stated the court, "he has no eyes and no brain, although he has an underdeveloped brain stem that supports minimal autonomic functioning. He has a cleft palate and cannot speak. He must be strapped into a wheelchair to sit, and he has no control of his bowels."[31] As in *Kurapati*, the court refused to allow the parents to sue for wrongful birth. (The court also disallowed the children from bringing claims for "wrongful life.") Of note is the concurring opinion by Judge Wintersheimer:

Any quality of life ethic favors the life of the healthy over the infirm, the able-bodied over the disabled and the intelligent over the mentally challenged. If logically extended, it could produce a culture that condones the extermination of the weak by the strong or the more powerful. The Nazi regime under Adolph [sic] Hitler is a not too distant reminder of this kind of eugenic approach. Unfortunately, such thoughts are not limited to foreign nations but can also be found in the writings of Justice Oliver Wendell Holmes in *Buck v. Bell* . . . , which approved of sterilization of the mentally incompetent. *Taylor, supra,* calls to our attention the influence that Hitler's experiments with sterilization had on the American eugenics movement. Eugenics espouses the reproduction of the fit over the unfit and discourages the birth of the unfit.[32]

In short, the judges in *Kurapati* and *Grubbs* did not want the machinery of the courts to contribute to what they felt was a eugenic goal, the desire of some parents to avoid giving birth to a child with severe disabilities. (Ironically, given their goal, they reached the wrong outcome. Arguably, parents are less likely to abort fetuses with disabilities if they can recover the costs of caring for the children through wrongful birth actions. On the other hand, physicians who had to compensate parents in wrongful birth cases might be more likely to encourage the parents to abort the fetuses in marginal cases.)

Another method by which the government involves itself in what might be considered neoeugenics is providing public funding for family-planning programs aimed at preventing the birth of children to poor mothers. According to the Alan Guttmacher Institute, the government in 2001 spent $1.26 billion on reversible contraceptive services and $95 million on sterilization services.[33] These funds are distributed through several programs: Medicaid, Title X, Title V (Maternal Child Block Grants), Social Services Block Grants, and TANF (Temporary Assistance for Needy Families).[34] These programs would not be particularly neoeugenic if they provided family-planning assistance to all socioeconomic groups, but virtually all of these funds are earmarked for the poor. As Jane Gilbert Mauldon writes, government-supported "family planning" efforts have been linked to eugenics in the past, and they arguably continue to pursue eugenic goals:

Eugenics seeks to improve social welfare by discouraging births of certain types, or births among certain social groups, and by encouraging other types of births. This line of thinking has a long history in the United States. Early in the twentieth century, explicitly eugenicist ideas, manifested as exhortations to educated women to have large families so as not to be out-bred by poor, immigrant or African American women, were commonly accepted among elites in the United States. More than thirty states had sterilization laws that provided for the involuntary sterilization of so-called socially unfit individuals, which could mean simply women who were poor and ill-educated. In 1958, Mississippi State Representative David H. Glass sponsored an unsuccessful bill to ". . . Discourage the Immorality of Unmarried Females by Providing for Sterilization of the Unwed Mother under Conditions of the Act; and for Related Purposes." Glass made clear the intended purpose of the act: "During the calendar year 1957, there were born out of wedlock in Mississippi, more than seven thousand Negro children . . . the Negro woman because of child welfare assistance [is] making it a business, in some cases, of giving birth to illegitimate children. . . . The purpose of my bill was to try to stop, slow down, such traffic at its source."

In some respects, contemporary policies still seem to reflect these elitist assumptions and concerns about "excess" childbearing among disadvantaged women. If programs were exclusively intended to assure recipients reproductive freedom, they might also include subsidized fertility services, not just contraception and sexually transmitted disease prevention. If they were entirely philanthropic programs intended to reduce unplanned childbearing across the board, they might have been designed to serve a larger group of people.[35]

Organizations that provide these federally funded family planning services are typically nonprofits, and therefore another way that the government facilitates their activities is by giving them a subsidy in the form of exempting them from federal and state taxes. (At the same time that it provides financial incentives for family planning, however, Congress prohibits the federal government from paying for abortions under Medicaid except in cases of rape or incest, or when a pregnant woman's life is endangered by a physical disorder, illness, or injury. Moreover, the United States will not provide aid to any foreign family planning organization that performs or promotes abortion. Clearly, many people do not regard abortion as an acceptable means of achieving eugenic goals.)

Another government tax policy that might be considered neoeugenic is federal child tax credits. Until statutory limits are reached, the more children the taxpaying family has, the larger the number of credits that it can claim. The policy might be regarded as eugenic in that, since only families with enough income to pay taxes benefit from the credit, the policy creates an incentive for better-off families to have more children.

Contrast child tax credits with so-called family caps under state welfare programs. Unlike child tax credits that encourage taxpaying families to produce more offspring, family caps apply to families on welfare, discouraging them from becoming larger by halting increases in welfare benefits once there are more than a certain number of children. Currently 24 states have some version of this policy.[36] Dorothy Roberts observes that "like birth control programs and reproductive punishments, contemporary welfare policies share features of eugenic thinking.... Of course, the current welfare family caps are not premised on notions of recipients' genetic inferiority. But, like eugenic programs of the past, they are seen as a way of ridding America of the burden poor people pose."[37]

In *Dandridge v. Williams,* the Supreme Court upheld family caps in the face of constitutional challenge.[38] The Court ruled that the state of Maryland did not violate the Equal Protection Clause of the Fourteenth Amendment because it had a reasonable basis for the policy, namely, "the state's legitimate interest in encouraging employment and in avoiding discrimination between welfare families and the families of the working poor."[39] Interestingly, the state itself had articulated an additional goal, "providing incentives for family planning,"[40] on which, without comment, the Court apparently declined to rely.

LEGALLY MANDATED PRACTICES

So far we have considered government toleration or subsidization of practices that can be considered neoeugenic. Additionally, there are some practices that the government currently mandates by law. All of them are in aid of negative eugenics objectives in that they discourage reproduction by undesirable segments of the population. One example is sterilization and other restrictions on reproduction employed ostensibly to deter the commission of future crimes. For example, California in 1996 enacted a law authorizing voluntary chemical or surgical castration for the first offense of child molesting and mandating castration as a condition of parole for repeat offenders.[41] Georgia, Florida, and Montana have adopted similar statutes. Chemical castration involves the use of Depo-Provera, a synthetic progesterone that decreases testosterone, causing a reduction in sex drive and an improved ability to control sexual fantasizing.[42] In another illustration, a judge conditioned parole for a convicted child abuser on surgical implantation of the contraceptive Norplant. The defendant challenged the requirement, but the case was dismissed as moot after other parole violations occurred.[43] Although the intent of these laws arguably is to prevent child molestation, they are eugenic in that they decrease the likelihood that child molesters will reproduce.

How can these statutes be constitutional when, as noted earlier, the Supreme Court invalidated the sterilization requirement in *Skinner*? The answer lies in the fact that the Court has never overturned its decision in *Buck v. Bell*, which upheld Virginia's sterilization program for inmates at its State Colony for Epileptics and Feeble Minded.[44] As Justice Douglas explained in *Skinner*, inmates were sterilized at the Virginia institution so that they might be allowed to leave it. As the Court stated in *Buck*, "So far as the operations enable those who otherwise must be kept confined to be returned to the world, and thus open the asylum to others, the equality aimed at will be more nearly reached."[45] In *Skinner*, Douglas pointed out, "there is no such saving feature."[46] The California, Georgia, Florida, and Montana laws likewise offer sterilization as an alternative to continued incarceration.

The criminal justice system deters reproduction not only by sex offenders but by all incarcerated criminals. In a broader sense, then, the entire criminal system is a negative neoeugenic program. Only five states (California, Mississippi, New Mexico, New York, and Washington) permit prisoners to have conjugal visits with their spouses, despite some evidence that such visits reduce the incidence of prison rape.[47]

In the early twentieth century, eugenic goals fueled restrictive immigration laws. Similar objectives arguably are associated with current efforts to limit immigration, including the plan to construct a fence along the border with Mexico. Although anti-immigration forces avoid directly linking immigration with eugenics, the connection between eugenics organizations like the Pioneer Fund and anti-immigration lobbying is too plain to dismiss.[48]

Another governmental action that could be considered neoeugenic was Congress's 2006 reversal of the federal executive order that had mandated coverage of Viagra under Medicaid.[49] The action was prompted by reports that some Medicaid programs had prescribed the drug for sex offenders. But it also could have the effect of discouraging the poor from reproducing.

After the FDA approved Norplant, a surgically implanted, long-lasting contraceptive, legislators in a number of states proposed linking Norplant use to eligibility for welfare benefits. In other words, women on welfare could not receive benefits if they continued to have children. Although no such requirement has been adopted, Norplant has been covered under all state Medicaid programs.

The state-mandated initiative that most overtly raises eugenics concerns, however, is newborn screening for nontreatable disorders. Newborn screening began in the 1960s after physician Robert Guthrie developed a screening test for PKU, an autosomal recessive metabolic disorder that can be treated effectively if detected soon after birth. Guthrie also pioneered a method for collecting and transporting on special filter paper the blood samples used for screening, known as "Guthrie cards," and he insisted that the collection and analysis of the samples be performed by state public health officials. Massachusetts adopted newborn screening on a voluntary basis in 1962, but after President Kennedy's Advisory Committee on Mental Retardation recommended mandatory screening, states began to enact newborn screening as a legal requirement. By 1973, newborn screening was compulsory in 43 states. Now it is universal.

The development of faster and cheaper technologies such as tandem mass spectrometry and microchip arrays enable programs to screen for far greater numbers of disorders, including many for which no readily effective treatments presently exist. Screening for these nontreatable disorders can be beneficial, in that it could spare families years of uncertainty once symptoms emerged; alert them to be on the watch for new discoveries that could provide their children with treatment; provide children with adjunctive if not curative interventions; and facilitate participation of the children in research on their disorders. Yet some public health advocates offer an additional rationale for screening for nontreatable disorders: that it can serve as a valuable tool in family planning. One recent article explains, for example: "Arguments for considering broader benefits from the early diagnosis that only newborn screening can provide include . . . knowledge on which to base reproductive decision-making years before a disease would be diagnosed for the affected child."[50]

Moreover, as noted above, newborn screening is mandatory. Only Maryland, Wyoming, and the District of Columbia currently seek parental consent for newborn screening. (Massachusetts recently has begun doing so in a pilot program.) In Maryland, the consent is for the total screening package; parents are not asked to consent to specific tests. Thirty-three states provide an exemption from screening if contrary to parents' religious beliefs, but it is up to the parents to assert the objection without being asked.

In addition, in a case called *Douglas County v. Anaya,* the Nebraska Supreme Court in 2005 rejected a constitutional challenge to newborn screening based on religious grounds. After a home birth, the Anayas refused to allow a blood sample to be taken and sued the public health department to block enforcement of the law, which contained no exceptions. The court refused to recognize a religious exemption, noting that "[t]he health and safety of the child are of particular concern."[51]

By focusing on the health and safety of the child, the decision makes the case seem no different than the numerous instances in which the courts have refused to allow parents such as Jehovah's Witnesses to withhold treatment from children for religious reasons. But courts override parental treatment decisions for religious reasons only when the denial of treatment would be fatal or would seriously impair the child's health. At least some newborn screening tests may not have such a direct connection to the child's welfare. Indeed, according to pediatric geneticist Jeffrey Botkin, of the 29 disorders for which the American College of Medical Genetics recommends screening, 12 are quite rare, occurring in less than 1 in 100,000 births.[52] In any event, the Supreme Court of Nebraska did not rest its decision in the *Anaya* case on the need for screening to protect the newborn's health alone. In addition, the court observed that mandatory screening was necessary in order to address "the potential social burdens created by children who are not identified and treated."

Newborn screening for treatable disorders is laudable. Screening for nontreatable disorders, however, is another matter. The screening is mandated by the state and is imposed on parents over their objection. Nontreatable disorders would be included in part in order to discourage parents from giving birth to additional children with those disorders. This will prevent these children from inheriting genes that will make them a burden on society. Viewed in this manner, the resemblance between this program and classic eugenic practices is, at the very least, disconcerting.

In her 2008 book *Heredity and Hope,* historian Ruth Schwartz Cowan defends medical genetics in general, and newborn screening in particular, from being attacked as eugenic. Their intention is not, she explains, the eugenic goal of preventing the genetically unfit from reproducing, but the exact opposite: to help people reproduce who suspect that they are at risk for having a genetically unfit child. According to Cowan, medical genetics facilitates this by allowing these people to confirm through genetic testing that they are not in fact at risk or, if testing shows that they are in fact at risk, by enabling them to use prenatal diagnosis to bring only genetically healthy fetuses to term.[53] Cowan is correct in that mandatory newborn screening does not necessarily aim at preventing reproduction by persons who may have undesirable genetic loads. But it does discourage the birth of children with undesirable mutations, with the result that our genetic stock will be improved. When this objective is extended to nontreatable disorders and accomplished through compulsory government programs, its neoeugenic pedigree is unmistakable, even if it is not intended.

THE LAW OF MODERN NEOEUGENICS

Clearly, laws and regulations that mandate or actively encourage neoeugenic practices, such as newborn screening for nontreatable disorders or family welfare caps, could be revised or repealed. But what about passive government policies? Given our current constitutional framework, could the law be changed to outlaw neoeugenic practices?

Many neoeugenic practices, including selective breeding, contraception, nontherapeutic abortion, and possibly also the right to employ assisted reproductive technologies, such as IVF and PGD, would be protected under the constitutional right to reproductive freedom. This right, incidentally, was first recognized by the Supreme Court in *Skinner v. Oklahoma,* the case striking down the Oklahoma law that authorized the sterilization of certain criminals, where the Court observed that "marriage and procreation are fundamental to the very existence and survival of the race."[54] As John Robertson and others have argued, the right of reproductive freedom arguably extends far enough to bar the law from interfering with a parental decision to prevent the birth of a child with disabilities and to promote the birth of a child with a favorable ratio of desirable traits.[55] A more difficult question is whether this right embraces the freedom to conceive a child intentionally with a disability, such as certain deaf couples are reported to desire.

Other neoeugenic practices are protected by the First Amendment. These include exhortative and religious activities such as private family planning educational initiatives and community-based genetic match-making programs. A scheme such as Project Prevention would not fall completely within the ambit of the First Amendment, however, since it involves action (paying women to be sterilized) as well as speech.

One approach for restricting neoeugenic practices might be to regulate health care professionals and provider organizations. There is no question, for example, that the state can regulate the activity of fertility doctors and clinics; as noted earlier, for instance, a federal law requires fertility clinics to disclose their success rates. At the same time, however, these laws must be constitutional. For example, the Supreme Court has invalidated laws prohibiting physicians from prescribing contraceptives[56] or performing abortions.[57] Therefore, the Court might strike down laws constraining IVF practitioners from assisting parents in making reproductive choices that may have neoeugenic consequences.

The government exercises broad regulatory authority over medical products as well as services. Thus it could restrict the marketing of genetic tests used in selective breeding, PGD, and newborn screening if it deemed them unsafe or ineffective. A recent case in the U.S. Court of Appeals for the District of Columbia, *Abigail Alliance v. von Eschenbach,* reaffirmed the government's power in the face of a challenge mounted by terminally ill cancer patients. The patients argued that they had a constitutionally protected right to obtain Phase I cancer drugs—that is, drugs that had completed only the first of the three stages of the testing required by the FDA for marketing approval. A previous opinion by a three-judge panel of the court had sided with the patients, but the full court reheard the case and reversed the decision on the ground that the right to self-preservation does not override the government's interest in protecting people from unsafe drugs.[58]

It remains to be seen, however, whether the government could exert its power to regulate the safety and efficacy of medical products to prevent them from being used to achieve neoeugenic goals. Once the FDA approves a drug or medical device, physicians are free to use them in any way they choose, subject only to liability for malpractice if their behavior is deemed unreasonable.[59] So even if the agency refused to approve a eugenic indication for a product—say, a label claim that a genetic test was useful in making reproductive decisions—FDA rules would allow a physician to perform the test for that purpose so long as the product had been approved for any use. If Congress or the FDA were to change this policy, however, the recent *Abigail* decision suggests that the new policy would be upheld as constitutional.

Aside from concerns about constitutionality, from a practical standpoint certain attempts to restrict neoeugenic practices would be unlikely to succeed. The private nature of reproductive decision making which lends it its cloak of constitutionality means that some enforcement activities would have to intrude into the patient-physician relationship, the internet, and the bedroom. Even if we were willing to permit such invasions, the sheer difficulty of monitoring these spaces would require enormous resources. Black markets offering banned genetic testing and assisted reproductive services are bound to spring up as they did when abortion was illegal. In 1962 alone, more than a million abortions were believed to have been performed, half by physicians.[60] Reproductive products and services that were illegal in the United States could be obtained abroad.[61]

FUTURE DEVELOPMENTS

As the understanding of genetic science expands, the ability to pursue neoeugenic objectives is bound to increase. Of greatest importance is the development of new tests to detect undesirable conditions and traits. As noted earlier, it is becoming possible to screen newborns for far greater numbers of genetic conditions and risk factors for disease. In June 2006, researchers in London announced that a new technique called preimplantation genetic haplotyping would increase the number of diseases or risk factors that can be detected by PGD testing from 200 to 6,000.[62] Similar testing could be performed on donor gametes and on fetuses in utero, and could be used by couples considering whether or not to reproduce.

In response to the threat of bioterrorism, some public health advocates believe that the nation's public health laws need to be revamped to give public health officials additional powers and discretion. With funding from the Robert Wood Johnson Foundation and the W. K. Kellogg Foundation, they have drafted a Model State Public Health Act.[63]

An indication of the scope of the powers that would be conferred on public health officials under this law can be obtained by considering its provisions for mandatory public health screening and testing. Under the act, a state or local public health agency may establish a compulsory screening program for any "conditions of public health importance that pose a significant risk or seriously threaten the public's health" [section 5-106(d)(1)]. The terms *significant risk* and *seriously threaten* are not defined in the act, but the term *condition of public health importance* is defined to mean "a disease, syndrome, symptom, injury, or other threat to health that is identifiable on an individual or community level and can reasonably be expected to lead to adverse health effects in the community" [section 1-102(6)]. Under the model act, then, a public health agency could make prenatal screening for genetic diseases and conditions mandatory for all pregnant women, so long as the agency determined that the diseases and conditions, if not detected in utero, posed a significant threat to health that can reasonably be expected to lead to adverse health effects in the community. Such a belief is not far-fetched. The American College of Obstetrics and Gynecology recently called for "routine" prenatal screening for Down syndrome.[64] Another provision of the model act states that a public health agency may make participation in such a screening program a condition "of participating in or receiving a service or privilege" [section 5-106(d)(2)]. Accordingly, women who refused to be screened could be denied health and welfare benefits.

Renewed interest in universal health insurance also could act as a spur to neoeugenic practices. Universal insurance provided by federal or state governments would give them greater control over the allocation of health care resources, which could lead to greater efforts to save money through eugenic measures. Even if the actual health insurers remained private companies, they would have greater incentives to try to reduce spending by refusing to cover genetic conditions that could have been, but were not, prevented. Legislation prohibiting genetic discrimination might not prohibit such practices; it is not yet clear that laws like the Genetic Information Nondiscrimination Act (GINA), for example, would necessarily do so. On the other hand, universal health insurance could help reverse health disparities that cause higher rates of morbidity, mortality, and infant mortality in poor and minority populations. By improving their chances of reproducing, this could militate against the achievement of neoeugenic objectives.

Although the development of human gene transfer technology—the ability to install new genetic instructions into existing cells—has been disappointing to date, a significant breakthrough could accelerate progress dramatically. This might lead to the development of genetic interventions that could prevent diseases or disorders if applied in utero or during IVF. In the same way that they mandate immunization, public health officials might require parents to employ these interventions, or they might provide incentives for doing so, such as public funding or tax breaks.

If genetic interventions took place early enough in the life cycle, the resulting changes in DNA would be incorporated into an individual's eggs or sperm and passed on to future generations. Societal pressure to use this technology to improve the germ line could become overwhelming. Future developments in genetics also hold out the possibility of manipulating DNA to enhance non-disease traits. Researchers have already created mice with genetically engineered improvements in strength[65] and cognitive ability.[66] The question for public policy would be whether to subsidize access to these technologies for those who cannot afford to pay for them. Failure to do so would be an implicit form of neoeugenics, since enhanced individuals would claim a greater share of societal resources, increasing their chances of reproductive success over genetic have-nots.

As we move forward 100 years after chapter 215 of the Indiana Acts of 1907, the question of how far the state may go in controlling reproductive decision making is bound to become more compelling. Some believe that the future of the genetic revolution holds the possibility of enabling people virtually to dictate the precise genetic endowment of their offspring. This would be nothing less than the ability to manage human evolution, the holy grail of eugenics. If this capability actually emerged, it would remain to be seen how the historical lessons of eugenics would inform its use.

NOTES

1. In 1921, the Supreme Court of Indiana held that law to be unconstitutional in *Williams v. Smith*, 131 N.E.2 (Ind. 1921). The Indiana legislature enacted a replacement sterilization law, *Ind. Acts 1927*, ch 241, which responded to the lack of due process in the first scheme.

2. Francis Galton, *Inquiries into Human Faculty and Its Development* (London: Macmillan, 1883), 17n1.

3. Francis Galton, "Eugenics: Its Definition, Scope, and Aims," *American Journal of Sociology* 1 (1904).

4. Michael J. Bamshad and Steve E. Olson, "Does Race Exist? An Overview/Genetics of Race," 289 *Scientific American* 78–85 (2003).

5. Dorothy E. Roberts, "Crime, Race, and Reproduction," 67 *Tulane Law Review* 1945, 1969 (1993).

6. Sonia Suter, "A Brave New World of Designer Babies?" 22 *Berkeley Tech. L. J.* 897, 968–69 (2007).

7. Daniel J. Kevles, "Eugenics," in 2 *Encyclopedia of Bioethics* 848 (3rd ed. 2004).

8. Troy Duster, *"Eugenics: Ethical Issues,"* in 2 *Encyclopedia of Bioethics* 854 (3rd ed. 2004) (emphasis in original).

9. *Skinner v. Oklahoma*, 316 U.S. 535 (1942).

10. *Relf v. Weinberger*, 372 F. Supp. 1196 (D.D.C. 1974), vacated as moot, 565 F.2d 722 (D.C. Cir. 1977).

11. Steve Weinberg, "Yao: China's Gift to Basketball," *Christian Science Monitor*, Jan. 3, 2006, at 13.

12. See, e.g., Alicia Ouellette et al., "Lessons across the Pond: Assisted Reproductive Technology in the United Kingdom and the United States," 31 *Am. J. L. and Med* 419 (2005).

13. Fertility Clinic Success Rate and Certification Act of 1992, Pub. L. No. 102-493, 106 Stat. 3146 (codified at 42 U.S.C. §§ 263a-1 to a-7 (2000)).

14. Ouellette, *supra* note 44, at 423.

15. 962 P.2d 67 (Utah 1998),

16. Id. at 72.

17. Id.

18. Id.

19. Lori B. Andrews, "A Conceptual Framework for Genetic Policy: Comparing the Medical, Public Health, and Fundamental Rights Models," 79 *Wash. U. L. Q.* 221, 245 (2001).

20. Beverly Merz, "Matchmaking Scheme Involves Tay-Sachs Problem," 258 *JAMA* 2636 (1987).

21. See Renee Chelian, "Remarks on the "CRACK" Program: Coercing Women's Reproductive Choices," 5 *J. L. Soc'y* 187 (2003).

22. *Molloy v. Meier*, 679 N.W.2d 711 (Minn. 2004).

23. Id. at 720.

24. *Taylor v. Kurapati*, 600 N.W.2d 670 (Mich. App. 1999).

25. Id. at 674.

26. Id. at 688.

27. Id.

28. Id.

29. Id. at 690.

30. *Grubbs v. Barbourville Family Health Ctr., P.S.C.,* 120 S.W.3d 682 (Ky. 2003).

31. Id. at 686.

32. Id. at 692.

33. Alan Guttmacher Institute, *Public Funding for Contraceptive, Sterilization, and Abortion Services* (2005) (http://www.guttmacher.org/pubs/fpfunding/tables.pdf).

34. Id., Table 1.

35. Jane Gilbert Mauldon, "Providing Subsidies and Incentives for Norplant, Sterilization, and Other Contraception: Allowing Economic Theory to Inform Ethical Analysis," 31 *J. L. Med. & Ethics* 351, 353–54 (2003).

36. See Janet Simmons, "Coercion in California: Eugenics Reconstituted in Welfare Reform, the Contracting of Reproductive Capacity, and Terms of Probation," 17 *Hastings Women's L. J.* 269, 277 (2006).

37. Dorothy Roberts, *Killing the Black Body: Race, Reproduction, and the Meaning of Liberty* (New York: Pantheon Books, 1997), 215–16.

38. 397 *U.S.* 471 (1970).

39. Id. at 486.

40. Id. at 484.

41. *Cal. Penal Code* sec. 645.

42. For more on chemical castration, see Avital Stadler, Comment, "California Injects New Life into an Old Idea: Taking a Shot at Recidivism, Chemical Castration, and the Constitution, " 46 *Emory L. J.* 1285 (1997).

43. Andrews, *supra* note 11, at 77, n.8.

44. *Buck v. Bell,* 274 U.S. 200 (1927).

45. Id. at 208 (quoted in *Skinner,* 316 U.S. at 542).

46. *Skinner,* 316 U.S. at 542

47. Rachel Wyatt, Note: "Male Rape in U.S. Prisons: Are Conjugal Visits the Answer?" 37 *Cas. W. Res. J. Int'l Law* 579 (2006).

48. See, e.g., Paul A. Lombardo, Commentary: "'The American Breed': Nazi Eugenics and the Origins of the Pioneer Fund," 65 *Alb. L. Rev.* 743, 750 (2002).

49. Cheryl G. Stolberg, "House Rejects Coverage of Impotence Pills," *New York Times,* June 25, 2005.

50. Duane Alexander (NIH) and Peter C. van Dyck (HRSA) 2006: "A Vision of the Future of Newborn Screening" 117 *Pediatrics* S350, 352.

51. *Douglas County v. Anaya,* 694 N.W.2d 601 (Neb. 2005).

52. Jeffrey Botkin, Testimony before the President's Council in Bioethics, Feb. 3, 2006 (http://www.bioethics.gov/transcripts/feb06/feb3full.html).

53. Ruth Schwartz Cowan, *Heredity and Hope: The Case for Genetic Screening* (Cambridge: Harvard University Press, 2008), 236–37.

54. *Skinner v. Oklahoma,* 316 U.S. 535, 541 (1942).

55. See, e.g., John A. Robertson, *Children of Choice: Freedom and the New Reproductive Technologies* (Princeton: Princeton University Press, 1994).

56. *Griswold v. Connecticut,* 381 U.S. 479 (1966).

57. *Roe v. Wade,* 410 U.S. 113 (1973).

58. *Abigail Alliance v. von Eschenbach,* 2007 U.S. App. *LEXIS* 18688 (Aug. 7, 2007).

59. David A. Kessler, "Regulating the Prescribing of Human Drugs for Nonapproved Uses under the Food, Drug, and Cosmetic Act," 15 *Harv. J. Legis.* 693 (1978).

60. See Zad Leavy and Jerome M. Kummer, "Criminal Abortion: Human Hardship and Unyielding Laws," 35 S. *Cal. L. Rev.* 123 (1962).

61. See Maxwell J. Mehlman and Kirsten M. Rabe, "Any DNA to Declare? Regulating Offshore Access to Genetic Enhancement," 28 *Am. J. L. and Med.* 179 (2002).

62. Ian Sample, "Embryo Test to Screen for 6,000 Inherited Diseases: Procedure Hailed as Major Advance for IVF Couples: Doctors Will Be Able to Test for Genetic Mutations," *Guardian,* June 19, 2006, 7.

63. Turning Point Model State Public Health Act (http://www.turningpoint program.org/Pages/pdfs/statute_mod/MSPHAfinal.pdf).

64. American College of Obstetrics and Gynecology News Release (http://www .acog.org/from_home/publications/press_releases/nr01-02-07-1.cfm).

65. BBC News, Aug. 24, 2004 (http://news.bbc.co.uk/2/hi/science/nature/ 3592976.stm).

66. Y. Tang, E. Shimizu, G. Dube., C. Rampon, G. Kerchner, M. Zhuo, G. Liu, and J. Tsien, "Genetic Enhancement of Learning and Memory in Mice," *Nature* 1999, 401: 63–69.

CONTRIBUTORS

ELOF AXEL CARLSON, Ph.D., is Distinguished Teaching Professor Emeritus in the Department of Biochemistry and Cell Biology at Stony Brook University in New York. His most recent book is *Neither Gods nor Beasts: How Science Is Changing Who We Think We Are.*

GREGORY MICHAEL DORR, Ph.D., was most recently Visiting Professor of Law, Jurisprudence, and Social Thought at Amherst College. He is author of *Segregation's Science: Eugenics and Virginia.*

MOLLY LADD-TAYLOR, Ph.D., is Associate Professor of History at York University in Toronto. She is author of *Mother-Work: Women, Child Welfare, and the State, 1890–1930.*

JASON S. LANTZER, Ph.D., is Adjunct Professor of History at Indiana University–Purdue University Indianapolis and Butler University and author of *Prohibition Is Here to Stay: The Reverend Edward S. Shumaker and the Dry Crusade in America.*

ANGELA LOGAN is a doctoral candidate in Philanthropic Studies at Indiana University. She currently serves as the Program Officer for Education at a private foundation in the southeastern United States.

PAUL A. LOMBARDO, Ph.D., J.D., is Professor of Law at Georgia State University in Atlanta. His most recent book is *Three Generations, No Imbeciles: Eugenics, the Supreme Court, and Buck v. Bell.*

EDWARD R. B. MCCABE, M.D., Ph.D., is Professor, Departments of Pediatrics and Human Genetics, David Geffen School of Medicine, and Department of Bioengineering, Henry Samuel School of Engineering and Applied Science; Mattel Executive Endowed Chair of Pediatrics Physician-in-Chief, Mattel Children's Hospital, and Co-Director, Center for Society and Genetics, all at the University of California at Los Angeles. He is author (with Linda L. McCabe) of *DNA: Promise and Peril.*

LINDA L. MCCABE, Ph.D., is Adjunct Professor, Departments of Human Genetics and Pediatrics, David Geffen School of Medicine, and Center for Society and Genetics, University of California at Los Angeles. She is author (with Edward R. B. McCabe) of *DNA: Promise and Peril.*

MAXWELL J. MEHLMAN, J.D., is the Arthur E. Petersilge Professor of Law and Director of the Law-Medicine Center, Case Western Reserve University School of Law, and Professor of Bioethics, Case Western Reserve University School of Medicine. His latest book is *The Price of Perfection: Individualism and Society in the Era of Biomedical Enhancement.*

JOHANNA SCHOEN, Ph.D., is Associate Professor of History at the University of Iowa. She wrote *Choice & Coercion: Birth Control, Sterilization, and Abortion in Public Health and Welfare.*

ALEXANDRA MINNA STERN, Ph.D., is the Zina Pitcher Collegiate Professor in the History of Medicine at the University of Michigan and author of *Eugenic Nation: Faults and Frontiers of Better Breeding in Modern America.*

INDEX

BIOETHICS AND THE HUMANITIES

ERIC M. MESLIN AND RICHARD B. MILLER, EDITORS

PAUL A. LOMBARDO is best known for his work on the history of the American eugenics movement, particularly the 1927 United States Supreme Court case of *Buck v. Bell,* which upheld state laws mandating eugenic sterilization of the so-called feebleminded and socially inadequate. He earned both his Ph.D. and J.D. at the University of Virginia, where he served on the faculty from 1990 until 2006. Since 2006 he has been Professor of Law at Georgia State University in Atlanta.